Contents

Handbook of Composite Fabrication

Editor: Güneri Akovali

RAPRA
TECHNOLOGY LTD.

Rapra Technology Limited

Shawbury, Shrewsbury, Shropshire SY4 4NR, United Kingdom
Telephone: +44 (0)1939 250383 Fax: +44 (0)1939 251118
http://www.rapra.net

First Published in English in 2001 by

Rapra Technology Limited

Shawbury, Shrewsbury, Shropshire, SY4 4NR, UK

©2001, Rapra Technology Limited

A catalogue record for this book is available from the British Library.

ISBN: 1-85957-263-4

Typeset by Rapra Technology Limited
Printed and bound by Polestar Scientifica, Exeter, UK

Preface

This book is designed as a handbook with some basic information on polymeric composite fabrication. It is prepared by considering beginners as well as technical personnel already working in this area. For this reason, it contains basic principles as well as an up-to-date information on various aspects, including applications.

The book has seven chapters. Each of these were prepared by a group of experts from different parts of the world. The first chapter, the introduction, gives some definitions and classifications with some depth on the matrices and reinforcements for all types of composite structures, including polymeric, metallic and ceramic matrix composites. More detailed information about constituent materials (matrices and reinforcements) is provided in chapter 2. Different processing techniques of polymer composites are discussed in the following chapters, briefly: open mould (in Chapter 3) and closed mould (matched die) processes, filament winding (Chapter 4) and pultrusion (Chapter 5). The book ends with the machining and joining processes commonly used for polymer composite structures.

I would like to thank each of the contributing chapter editors for preparing such a fine work, for being timely and cooperative. My thanks are also due to the commissioning editors Ms Frances Powers and Dr Arshad Makhdum of Rapra Technology Ltd for their ever-encouraging efforts and unceasing supports given during the preparation of the book. I should also like to extend my gratitude to Ms Claire Griffiths and Mr Steve Barnfield for their skillful typesetting and desk-top publishing of the text.

Finally, I should admit that, I greatly enjoyed editing this book. I hope the readers will enjoy using it at least as much, benefit from it and keep it as a valuable source book.

Professor Güneri Akovali, Editor

5 June 2001
Ankara

1

Acknowledgement

Tiğmur Akgül would like to thank Marcel Dekker Incorporated for permission to reprint material from M.K. Mallick, Fiber Reinforced Composites, Materials Manufacturing, and Design, 1988, p.332-335, ©1988, which appears on p.78-80 of this book.

1 Introduction

G. Akovali and N. Uyanik

1.1 Definition and classifications

A composite, in general, is defined as a combination of two or more components differing in form or composition on a macroscale [1], with two or more distinct phases having recognisable interfaces between them [1, 2].

Proper combination of materials into composites gives rise to properties which transcend those of the constituents, as a result of the principle of combined action. Materials of biological origin are generally composites. Bone, for instance, achieves its combination of lightness and strength by combining crystals of apatite (a compound of calcium) with fibres of the protein collagen, whereas wood contains cellulose fibres surrounded by lignin and hemicellulose. Crushed rock aggregate used in concrete produces a composite structure, which reduces the cost and helps to improve the compressive strength. Structural weight savings (while retaining the reliability and strength), are achieved for aerospace, rocket applications etc. by the use of composite materials.

Composites are produced to optimise material properties, mechanical (mainly strength), chemical and/or physical properties. In the latter, optimisation of thermal (thermal expansion/thermal conduction, specific heat, softening and melting points) as well as electrical (electrical conductivity/electrical permittivity, dielectric loss), as well as optical and acoustical properties can be noted. Since the early 1960s, there has been an increasing demand for materials that are stiffer and stronger, yet lighter in aeronautic, energy, civil engineering and in various structural applications. Unfortunately, no monolithic engineering material available is able to satisfy them. This need and demand certainly led to the concept of combining different materials in an integral composite structure.

Composites usually consist of a reinforcing material embedded in a matrix (binder). The effective method to increase the strength and to improve overall properties is to incorporate dispersed phases into the matrix, which can be an engineering material such as ceramic, metal or polymer. Hence, ceramic matrix composites, metal matrix composites (MMC) or polymer matrix composites (PMC)—or ceramic/metal/polymer composites—, carbon matrix composites (CMC) or even hybrid composites are obtained. In a composite, matrices, in general, are of low modulus, while reinforcing elements are typically 50 times stronger and 20–150 times stiffer. MMC and CMC structures are developed to provide rather high temperature applications (> 316 °C), where PMC are usually inadequate. Furthermore, since metals are more conductive (electrically and thermally), MMCs are also used in heat dissipation/electronic transmission applications. Each matrix type has a different impact on the processing technique.

Composites are usually used for their structural properties where the most commonly employed reinforcing component is in particulate or fibrous form and hence the definition above can be restricted to such systems that contain a continuous/discontinuous fibre or particle reinforcement, all in a continuous supporting core phase, the matrix. A reinforcement phase usually exists with substantial volume fractions (10% or more). Hence, three common types of composites can be described as: (a) particle strengthened, (b) discontinuous fibre reinforced and/or (c) continuous fibre reinforced composites; depending on the size and/or aspect ratio and volume fraction(s) of reinforcing phase(s). In these, the function of each component can be different: in particle-strengthened composites, the matrix bears the main load and small dispersed particles obstruct the motion of dislocations in the matrix; and the load is distributed between the matrix and particles. In fibre reinforced composites (FRC), the fibres bear the main load and the function of the matrix is confined mainly to load distribution and its transfer to the fibres. In addition to these types of composites, one should also note the existence of another group of composite system, laminar composites (or simply laminates); where reinforcing agents are in the form of sheets bonded together and are often impregnated with more than one continuous phase in the system [3]. Feldman [4], narrows the broad classification for composites given above simply into two groups: macrocomposites and microcomposites. This classification depends on whether there are one or more dispersed distinguishable phases (i.e., each larger than 1 mm) and on whether there is more than one continuous phase present (for macrocomposites), or whether all dispersed phases are between 10–1000 nm in size and there is only one continuous phase (for microcomposites). If the size of reinforcing component of a microcomposite is in the form of 'quantum dots' (i.e., being much smaller than 25 nm) then a new specific term is assigned to it by calling it a nanocomposite. The term flexible composite is used to identify composites based on elastomeric polymers where the usable range of deformation is much larger than conventional thermoplastic or thermosetting composites [5].

High tensile and/or high modulus fibres (such as carbon, boron, silicon carbide and alumina) emerged in the 1970s, and were used to reinforce high-performance polymer, metal or ceramic matrices. A new group of advanced composite materials (ACM) [6], were then developed which were, in general, extremely strong and stiff. The matrix is one of the key factors to reach to proper ACM structures. To have advanced material properties, firstly densities of matrices should be as small as possible with as high as application temperatures for the material. There usually exists a relationship between density and service temperature for different materials, as presented in **Figure 1.1**. As the figure shows, titanium and steel are not considered as advanced because of their relatively high densities. The arrow in the figure indicates the trend for advanced materials which has its peak at high application temperatures and low densities (two important criteria for the next generation spacecrafts being lighter and faster); hence silicone carbide ceramics and carbon-carbon composites are expected to be examples of such advanced materials.

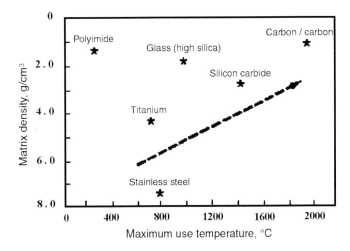

Figure 1.1 Density versus maximum use temperature for some materials

The unique properties of ACM make them special for different applications. For example, advanced polymer composites are not only lightweight, but also offer excellent strength, stiffness and design versatility, which are particularly important in aerospace applications. Some speciality polymers also have good chemical resistance and dielectric strength, like aramids, which gave rise to their use as electrical insulators in generators and transformers. Advanced ceramic materials provide a unique combination of high-temperature resistances, in addition to excellent wear and corrosion resistance and dimensional stability which are particularly important for parts which will be subjected to wear and cutting tools. Advanced metal composites and alloys obtained by rapid cooling of the melt have better strength and better electrical properties as well as improved corrosion resistance and enhanced magnetic properties.

The interfaces and interphases between different components in the composite, which is the boundary surface with a discontinuity, has a vital importance in determining the structural properties of the composite. The interfaces and interphases are expected to be proper, i.e., the interaction and adhesion between the components should be at the optimum level [7] in order to distribute the load that is borne by the composite evenly.

Finally, one should mention the three main factors that may affect the competition between composites and traditional engineering materials, both with similar mechanical properties: they are the cost, the reliability and the degree of complexities involved. The cost barrier in these is usually overcome by mass production and the degree of complexity is certainly more critical for composites which is due to the unisotropy existing (at least on microscale) for these structures. It is known that, in composites, thermoelastic properties as well as strength and failure modes have strong directional dependencies.

1.2 Structure of the matrix

The matrix usually comprises 30%–40% of composite structure. It has a number of functions:

a) it binds the components together and determines the thermo-mechanical stability of the composite,

b) it protects the reinforcements from wear/abrasion and environment,

c) it helps to distribute the applied load by acting as a stress-transfer medium,

d) it provides durability, interlaminar toughness and shear/compressive/transverse strengths to the system in general, and,

e) it maintains the desired fibre orientations and spacings in specific structures.

As regards the toughness and strength of the composite, the role of the matrix is more subtle and complex than that of the reinforcements involved. Most of the reinforcing components like glass, graphite and boron fibres, are all linear elastic and brittle solids, and whenever the stress on them is sufficient to cause unstable flaw growth, they fail catastrophically. And although both reinforcing component and the matrix are brittle, their combination can produce a material that is quite tough (at least much tougher than either of both components alone) via a synergism achieved by a combination of mechanisms that tends to keep cracks and flaws small, isolated and that dissipates mechanical energy effectively.

In the following sections, only brief definitions and some basic information in relation to matrix materials will be given. Some more detailed information for these will be presented in Chapter 2.

1.2.1 MMC

Metals are chosen as the matrix material in MMC structures mainly because of the following characteristics:

a) they have higher application temperature ranges,

b) they have higher transverse stiffnesses and strengths,

c) in general, they have high toughness values,

d) when present in metal matrices, the moisture effects and the danger of flammability are absolutely absent and they have high radiation resistances,

e) they have high electric and thermal conductivities,

f) MMC have higher strength-to-density, stiffness-to-density ratios, as well as better fatigue resistances, lower coefficients of thermal expansion (CTE) and better wear resistances as compared with monolithic metals, and

g) they can be fabricated with conventional metal working equipment.

However, on the negative side, the following disadvantages can also be mentioned for MMC:

a) most metals are heavy (titanium and stainless steel are not considered in the advanced group due to their high densities, **Figure 1.1**)

b) metals are susceptible to interfacial degradation at the reinforcement and matrix interface and are susceptible to corrosion [2], and

c) MMC usually have high material and fabrication costs and the related composite technology is not yet well matured.

Some typical properties of some metals and some of their alloys are listed in **Table 1.1** and the typical engineering stress-strain curves of steel and aluminium (with large plastic strains) are presented in **Figure 1.2**.

The most common metals employed in MMC are aluminium, copper and magnesium. Within these, aluminium has an application temperature at and above 300 °C and its alloys are normally used with boron or borsic filaments (silicon carbide coated boron), and 6061 aluminium (which provides a good combination of strength, toughness and corrosion resistance) is used more frequently than either 2024 (provides the highest

	Density ρ (g cm^{-3})	Young's Modulus (GPa)	Yield Strength (MPa)	Tensile Strength (MPa)
Table 1.1 Some typical properties of some metals and their alloys				
Pure metals				
Aluminium	2.7	70	40	200
Copper	8.9	120	60	400
Nickel	8.9	210	70	400
Ti-6A1-4V		110	900	1000
Aluminium alloys				
(high strength/low strength)		70	100–380	250–480
Stainless steel (304)		195	240	365
Plain carbon steel	7.9	210	250	420
Ti-6Al-4V: titanium-aluminium-vanadium				

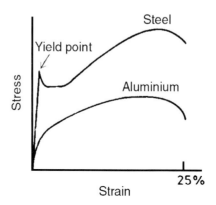

Figure 1.2 Stress-strain curves of aluminium and steel both with large plastic strains

strength) or 1100 (pure aluminium with high charpy impact resistance). Titanium can be used at 800 °C, usually as a matrix for borsic fibres [8]. Magnesium alloys are usually used with graphite reinforcements (with longitudinal tensile strength and moduli values of 177 MPa and 90 GPa for the composite, respectively. For the monolithic magnesium, tensile strength and moduli values are 65 MPa and 45 GPa, respectively. Magnesium matrices with boron fibres exhibit excellent interfacial bond and outstanding load redistribution characteristics.

Copper MMC are particularly of interest for materials with high thermal conductivities, e.g., for the need of an actively-cooled skin structure for the National Aerospace Plane of the USA (NASP) [9]. The high density and limited upper-application-temperature values of monolithic copper can be compensated by the introduction of graphite fibres, which give rise to MMC with reduced density, increased stiffness, increased application temperatures and improved thermal conductivities.

Discontinuously dispersed niobium (particle) reinforced copper composites are found to maintain the high thermal conductivity of the copper matrix with improved tensile strengths.

For advanced gas turbine engine component applications, aluminide matrices (i.e., FeAl, FeCrAlY or NiAl) are usually used, and are also called intermetallics. In general, they have excellent oxidation resistance, low density and high melting temperature. To have successful composites of intermetallics, the reinforcing fibres are expected to provide both toughening and strengthening to the system and for this, they must be chemically compatible with the matrix and should have similar CTE with the matrix. For example, Al_2O_3 fibres tested are found to create brittle and poorly-bonded reinforcements, while molybdenum and tungsten fibres, in general, give rise to a ductile and strongly-bonded composite [8].

In principle, all metals exhibit degradation of properties at very high temperatures, hence there is a thermal limitation in use even for the MMC as well.

1.2.2 Ceramic matrix composite

Ceramics are composed of metallic and non-metallic elements. Ceramics are chosen as the matrix material mainly because of following facts:

a) they have a very high application temperature range (>2000 °C), hence they provide advanced heat engine applications,

b) they have low densities, and

c) they usually have very high elastic modulus values.

The major disadvantage to ceramic matrix materials is their brittleness, which makes them easily susceptible to flaws. Besides being brittle, they usually lack uniformity in properties and have low thermal and mechanical shock resistances, as well as low tensile strengths. Hence the major disadvantage of ceramics is their brittleness—existence of even minor surface flaws, scratches or internal defects (pores, microcracks) can result in disaster. In fact, one of the driving forces in this field is the production of tough ceramic materials.

Commonly used ceramic matrix materials can be categorised in the following four main groups:

a)   glass ceramics (such as lithium aluminosilicate),

b)   oxides [such as alumina and mullite ($Al_6Si_2O_{13}$)],

c)   nitrides (such as silicone nitride, Si_3N_4), and

d)   carbides (such as silicone carbide, SiC).

Some authors [2, 9] consider carbon to be categorised as the fourth group.

Silicone nitride matrices are specifically used for the production of ceramic matrix composite systems where strong, tough, oxidation resistant and very high temperature/ high heat flux resistant materials are needed, e.g., for advanced heat engines. For these kind of applications, high temperature resistant fibres are being employed (30% aligned SiC, i.e., 'SiCS6' fibres) in ceramic matrices.

Table 1.2 Typical examples of some ceramic matrices				
	Density ρ (g cm^{-3})	Young's Modulus (GPa)	Yield Strength (MPa)	CTE (10^{-6} K^{-1})
Borosilicate glass	2.2	60	100	3.5
Soda glass	2.5	60	100	8.9
Mullite		143	83	5.3
MgO	3.6	210–300	97–130	13.8
Si_3N_4	3.2	310	410	2.25–2.87
Al_2O_3	3.9–4.0	360–400	250–300	8.5
SiC	3.2	400–440	310	4.8
Glass – ceramics:				
Lithium aluminosilicate	2.0	100	100–150	1.5
Magnesium aluminosilicate	2.6–2.8	120	100–170	2.5–5.5

Unidirectional and two dimensional $SiCS_6$/silicone nitride laminates are shown to exhibit high specific stiffness and strengths in addition to high toughnesses, notch insensitivity, creep resistance and thermal shock resistances up to 1400 °C. Also, significant matrix strength enhancements are predicted for this system with SiC fibres less than 40 mm in diameter [10]. In addition, gas turbine engines of supersonic aircrafts opened up the possibilities of use of ceramic matrix composites composed of SiC both as matrix and reinforcing fibres, which led to composite systems with high strength, high toughness, high oxidation resistance and high thermal conductivities for structural applications at temperatures above 1400 °C [11]. Also, the use of 'reaction-formed' silicone carbide reinforcement is claimed to have certain advantages. Properties of some ceramics are given in **Table 1.2**.

1.2.3 CMC

Carbon, as the matrix material, is usually used with carbon fibres in composite systems. The main advantage of the carbon matrix and its carbon/carbon (C/C) composite is its resistance to high temperatures in excess of 2200 °C, and the fact that it gains extra strength at elevated temperatures.

C/C composites have high strength-to-weight and high stiffness-to-weight values, have high dimensional stabilities and high resistance to fatigue. Hence, C/C composites are used in structural and non-structural applications where temperatures involved are high (e.g., on the leading edge and on the rocket engine nozzles of the NASA space shuttle where there are very high temperatures observed during flight and re-entry, as well as in the brake system of airplanes). In addition to the extensive use of this composite for very high temperature applications, specific applications in the medical field are also being developed; since carbon is chemically and biologically inert, it can be kept sterile inside the human body and can be used as prosthetic devices. Additionally, this composite is considered to be very similar to bone and has been used for some time in replacement hips or joints. Many internal or surgical implant devices are already being produced, preferably from carbon composites, because they also exhibit greater corrosion resistance and chemical resistance than stainless steel or selective metal alloys, in addition to their inherent satisfactory fatigue and toughness characteristics. Conversely, the main disadvantage of using C/C composite is the cost, i.e., the cost of the materials used and the cost of the fabrication of the composite.

1.2.4 PMC

Polymers are mostly organic compounds based on carbon, hydrogen and other non-metallic elements. PMC are the most developed composite materials group and they have found widespread applications. PMC can be easily fabricated into any large complex shape, which is an advantage.

In PMC applications, thermosetting or thermoplastic polymers can be used as the matrix component. PMC (also called reinforced plastics) are, in general, a synergistic combination of high performance fibres and matrices. In these systems, the fibre provides the high strengths and moduli while the polymer matrix spreads the load and helps resistance to weathering and to corrosion. Thus, in PMC, strength is almost directly proportional to the basic fibre strength and it can be further improved at the expense of stiffness. Optimisation of stiffness and fibre strength is still one of the unresolved main objectives and is under serious consideration. In some cases, mainly due to the differences in flexibilities between reinforcing fibres intra- and inter-fibrillar amorphous zones, severe shear stresses can result in the system, eventually leading to a fatigue crack.

Thermoplastic PMC soften upon heating at the characteristic glass transition temperature (T_g) of the polymer which, usually, are not too high (upwards of 220 °C). Hence, for thermoplastic PMC:

a) the application temperatures are rather limited,

b) they can be readily processed by use of conventional plastics processing techniques, such as injection moulding, extrusion and blow moulding, and

c) they can be reshaped easily with heat and pressure, in addition to the fact that they offer the potential for the higher toughness and the low cost-high volume processing of composite structures, and

d) one of the main disadvantages of considering thermoplastics as a matrix material is their rather large CTE values, which may lead to a mismatch in their composites and their sensitivities towards environmental-mostly hygrothermal-effects (i.e., absorption of moisture causes swelling as well as reduction in T_g, leading to severe internal stresses in the composite structure).

The most commonly used thermoplastic matrix materials are polyolefinics (polyethylene, polypropylene), vinylic polymers (polyvinyl chloride (PVC)), polyamides (PA), polyacetals, polyphenylenes (polyphenylene sulphide (PPS)), polysulphone and polyetheretherketone (PEEK). Some of their characteristic properties are presented in **Table 1.3**.

Conversely, thermosetting PMC are crosslinked and shaped during the final fabrication step, after which they do not soften by heating. They have a covalently-bonded, insoluble and infusible three-dimensional network structure. In order to promote processability, thermosetting resins are typically available in the special B-stage, which refers to a partially cured and usually vitrified system below the gel point. The combination of reinforcement and B-staged resin, usually in sheet form of approximately 1 mm thick, is termed prepreg (an abbreviation for pre-impregnation). The final, fully cured resin is referred to as the C

Table 1.3 Typical properties of some thermoplastic and thermosetting matrices							
Property	Epoxy	Thermosetting polyimides	PEEK	Polyamide-imide	Polyether imide	Poly-sulphone	PPS
Density (g cm⁻³)	1.15–1.4	1.43–1.46	1.30	1.38		1.25	1.32
Elastic Modulus (GPa)	2.8–4.2	3.2					
Flexural Modulus (MPa)	15–35	35	40	50	35	28	40
Tensile Strength (MPa)	35–130	55–120	92	95	105	75	70
Compressive Strength (MPa)	140	187					
CTE (10^{-5} °C)	4.5–11	5–9		6.3	5.6	9.4–10	9.9
Thermal Conductivity (W m⁻¹ K⁻¹)	0.17–0.2	0.36					
Water Absorption (24 h, %)	0.1	0.3	0.1	0.3	0.25	0.2	0.2
T_g (°C)	130–250	370	310				
Continuous Service Temperature (°C)	25–85	260–300			170	175–190	260

stage, and corresponds to the final fabricated form. High performance PMC usually involve stacks of appropriately-oriented prepregs subsequently cured in an autoclave.

Most of the potential disadvantages of using thermoset matrices are common with those of thermoplastics, although heat resistances are much higher and there is no softening point involved in the case of thermosets. In brief, the application temperature ranges for both are still limited, both are susceptible to environmental degradation due to radiation/ moisture and even atomic oxygen (i.e., in the space), they have rather low tranverse strengths and there may be very high residual stresses due to the mismatch in CTE between reinforcement and the matrix.

The most common thermoset polymer matrix materials are polyesters (unsaturated), epoxies and polyimides. Although information about these will be given with some depth in Chapter 2, the basic properties are considered briefly in this part. Polyesters are extensively used with glass fibres, they are inexpensive, are somewhat resistant to environmental exposure and are lightweight with useful temperatures up to 100 °C. They are the most widely employed class of thermosetting resins used for automotive, various construction and in general for most of the non-aerospace applications. Their poor impact and hot/wet mechanical properties, limited shelf life and high curing shrinkages avoid their high performance applications. Epoxies are more expensive than polyesters and have lower shrinkage on curing. They show good hot/wet strength, excellent mechanical properties, dimensional stability, good adhesion to a variety of reinforcements and a better moisture resistance. Also, epoxies maximum application temperature is slightly higher (175 °C). A large number of different types and different formulations are available for epoxies. Most of the high performance PMC have epoxies as matrices. Polyimides are difficult to fabricate, however, since they have a much higher application temperature (300 °C).

A convenient polymer matrix composite can be converted into a ceramic matrix composite (i.e., a silicone oxycarbamide ceramic composite which can be produced by pyrolysis followed by long-term oxidation processes from a 'pre-ceramic PC', the details of which are not disclosed [12]).

1.2.5 Hybrid composite material (HCM)

HCM represent the newest group of various composites where more than one type of fibre is used to increase cost-performance effectiveness. That is, in a composite system reinforced with carbon fibre, the cost can be minimised by reducing its content while maximising the performance by optimal placement and orientation of the fibre. Aramid reinforced aluminium laminate (ARALL) is another example of such a composite, consisting of high strength aluminium alloy sheets interleafed with layers of aramid fibre reinforced adhesive.

Other HCM include: nanocomposites [11], functionally gradient materials [13], hymats (hybrid materials) [14], interpenetrating polymer networks [15], and liquid crystal polymers [16].

1.3 Reinforcement forms

In the broadest sense, composites are the result of a combination of substantial volume fractions of high strength, high stiffness reinforcing components with a lower modulus matrix. The properties of composites are a function of the properties of the constituent phases, their relative amounts, as well as of the geometry of the dispersed phase (i.e., shape of the reinforcing component and size, their distribution and orientation).

As mentioned previously, the reinforcing component in a composite structure can be discontinuous (either in the form of dispersions/particles, flakes, whiskers, discontinuous short fibres with different aspect ratios) or continuous (long fibres and sheets); although the most commonly employed reinforcing component is in particulate or in fibrous form. Usually, materials in fibre form are stronger and stiffer than in any other form, which is the main reason for the overwhelming attraction for fibrous reinforcements.

1.3.1 Geometry of reinforcing materials

In general, composites can simply be classified into three groups as particle reinforced, fibre reinforced and structural composites [17], and at least two sub-divisions exist for each in this kind of classification:

a) the dispersed phase for particle (or particulate) reinforced composites is equiaxed (particle dimensions are approximately the same in all directions) and include those that are spherical, rod, flake-like and with other shapes of components with roughly equal axes. It is to be noted that this characterisation may not be complete if additional characteristics such as porosity are not considered too. Fillers in filled systems where there may be no reinforcement at all, and where particles are used to extend the material for cost-reduction purposes, cannot be considered as particulate reinforced. The same applies for particles added for non-structural purposes, e.g., flame retarders, etc.

b) fibre reinforced composites, where the dispersed phase is fibrous with a larger length-to-diameter ratio;

c) while structural composites are combinations of composites and homogeneous materials (i.e., laminates and sandwich panels), there is always the possibility of a change of shape (and even size) of particulates and sizes of fibres in the first two of these groups depending on the type of processing employed during the processing stage. Usually, materials in fibre form are much stronger and stiffer than any other form and mainly for this reason, there usually is an overhelming attraction for the fibre reinforcements.

1.3.1.1 Particulate Reinforced Composite Systems

Particles (or particulates) can have various effects on the matrix material depending on the properties of two components: ductile particles added to a brittle matrix in general increase the toughness (as cracks have difficulty passing through them), e.g., rubber modified polystyrene (HIPS) [18], and epoxy systems [19]. Conversely, if particles are hard and stiff, (i.e., with high moduli values) and are used in a ductile matrix, an increase in strength and stiffness usually results, (e.g., carbon black added to rubber). Unfortunately, hard particles generally decrease the fracture toughness of a ductile matrix, thereby limiting the applicability of this type of composite system. However, if other desirable properties can outweigh the disadvantages of limited toughness, they may be useful and preferred, e.g., the high hardness and wear resistances gained by the use of cemented carbides. The reduction in toughness for such systems usually is size and concentration-dependent and its magnitude may not be too high and if the size of hard particles is quite small and they are in limited quantity, the negative effect involved can easily be overcome. For instance, in a metal matrix (such as aluminium) an additional strengthening effect can be achieved by dispersion hardening, obtained by sintering the metal powder with ceramic particles, such as aluminium oxide, of 0.1 mm size with loadings between 1%–15%. Tungsten, with sufficient creep resistance values for use in light-bulb filaments, can be dispersion hardened with small amounts of oxide ceramics like thorium oxide, aluminium oxide, etc.

Particle reinforced composite systems can be considered in two sub-classes: 'large particle' and 'dispersion' strengthened composites, by basing on the reinforcement or strengthening mechanisms. The term 'large' is used to denote that particle-matrix interactions can not be treated on atomic or molecular level and 'continuum mechanics' is used. For composites reinforced by large particles, the reinforcing component is usually harder and stiffer than the matrix and they tend to restrain movement of the matrix. The matrix transfers some of the applied stress to the particles. The efficiency of reinforcement depends strongly on interaction at the particle-matrix interface. Although particle sizes are rather large (> 0.1 mm) and exist with a variety of different geometries, they have approximately the same dimension in all directions (equiaxed).

For effective reinforcement, the particles need to be evenly distributed and not too large. Mechanical properties are dependent on volume fractions of both components and are enhanced by an increase of particulate content.

Generally, 'the rule of mixture' equations are used to predict the elastic modulus of a composite under iso-strain conditions for the upper bound (Equation 1.1).

$$E_c = (E_m V_m) + (E_p V_p) \qquad (1.1)$$

as well as lower bound for iso-stress condition (Equation.1.2),

$$(E_c')^{-1} = (E_m V_m)^{-1} + (E_p V_p)^{-1} \qquad (1.2)$$

where E, E′ and V are moduli (elastic and transverse) and volume fraction, respectively; and the subscripts c, m, p represent the composite, the matrix and the particulate components, respectively. Large particle reinforced composite systems are utilised with all three types of materials (metals, ceramics and polymers). Concrete is a common large particle strengthened composite where both matrix and particulate phases are ceramic materials.

In dispersion strengthened composites, dispersed particle sizes are very small (between 0.01–0.1 mm) and the matrix bears the main load while the small dispersed particles obstruct and hinder the motion of dislocations in the matrix, so that the plastic deformations are restricted and the yield and tensile strengths (as well as hardness), improve particle-matrix interactions which occur at the molecular (or atomic) level. Many metal and metal alloys can be reinforced and hardened by the uniform dispersion of several percent of very hard, inert and fine particles as dispersions.

Elastomers and rubbers are usually reinforced with various particulate components, carbon black being the most common one. Carbon black, consisting of very small (with diameters 0.02–0.05 mm), essentially spherical particles, is a very efficient reinforcing component for vulcanised rubber; it enhances tensile strength, toughness, tear and abrasion resistances when used with 15%–30% by volume.

1.3.1.2 Fibre reinforced composite (FRC) systems

A FRC is the most technologically important group, where the dispersed phase is in the form of a fibre. In these systems, fibres bear the main load and the function of the matrix is confined mainly to load distribution and its transfer to the fibres as well as to hold them in place. In general, design goals for FRC include stiffness and/or high strength, both on a weight basis, which are termed as specific modulus (the ratio of elasticity modulus to specific gravity) and specific strength (the ratio of tensile strength to specific gravity), respectively.

The word fibre, means a single, continuous material whose length is at least 200 times its width or diameter and filaments are endless or continuous fibres [20]. Whereas, the word whisker is used for 'single crystal', defect-free metal fibres consisting of rather short, discontinuous fibres of 'polygonal cross sections', e.g., whiskers of copper, graphite, silicone carbide etc. Whiskers of oxide, carbide and nitride are of primary engineering interest. There are also multi-phase fibres, which are fibres formed specifically on the surface of an another kind of fibre, e.g., boron nitride or silicone carbide formed on the surface of a very fine wire tungsten substrate.

Yarn is a continuous bundle or strand of fibres. In the fibre world, both the size and density of fibre is expressed in a combined one word: denier (abbreviation, den), which is the weight (in grams) per 9000 m of fibre.

There are two different classes of fibres: natural (fibres from mineral, plant and animal sources) and synthetic (man-made fibres). Within these two classes, synthetics are usually more uniform in size, are more economical to use and behave in a more predictable manner. For engineering applications the most commonly employed significant fibres are glass fibres, metallic fibres and organically-derived synthetic fibres. Most strong and stiff fibres (e.g., ceramic fibres of glass, graphite-carbon, boron carbide and silicone carbide) are usually difficult to use as structural materials in bulk. However, embedding such materials in a ductile matrix (such as a polymer or metal) enables them to behave as a stronger, stiffer, tougher material. As explained previously, proper adhesion between fibres and matrix is very important since this enables the matrix to carry the stress from one fibre to another (in case a fibre breaks).

Fibre reinforcement can be carried out by using either discontinuous or continuous fibres. Various fibre forms are presented in Chapter 2 in some detail.

References

1. H.E. Peply in *Engineered Materials Handbook*, Ed, T. J. Reinhart, ASM International, Ohio, USA, 1987, Volume 2, Chapter 2.3.

2. C.T. Herakovich in *Mechanics of Fibrous Composites*, J. Wiley and Sons Inc., New York, 1998.

3. L. Hollaway in *Polymer Composites for Civil and Structural Engineering*, Chapman and Hall Inc., Glasgow, 1993.

4. D. Feldman in *Polymeric Building Materials*, Elsevier Applied Science, London, 1989.

5. S.Y. Lou and T.W. Chou in *Composite Applications – The Role of Matrix, Fibre and Interface*, Eds., T.L. Vigo and B.J. Kinzig, VCH Publishers Incorporated, New York, USA, 1992, Chapter 2.

6. L.A. Pilato and M.J. Michno, in *Advanced Composite Materials*, Springer-Verlag, Berlin, 1994.

7. L.H. Sharpe in *The Interfacial Interactions in Polymeric Composites*, Ed., G. Akovali, Kluwer Academic Publishers, Dordrecht, 1993, NATO Science Series, Volume 230.

8. R.R. Bowman, R.D. Noebe and J. Doychak, Presented at 4th Annual HITEMP, CD-91, Cleveland, Ohio, USA, 1991, 57073, 43-1.

9. M.M. Schwartz in *Composite Materials—Properties, Nondestructive Testing and Repair*, Prentice Hall, New Jersey, 1997.

10. R.T. Bhatt, Presented at the 4th Annual HITEMP, CD-91, Cleveland, Ohio, 1991, 57179, 67-1.

11. S. Komarneni, *Journal of Materials Chemistry*, 1992, 2, 12. 1219.

12. *Advanced Composites Bulletin*, 1999, April, 3.

13. J.S. Moya and A.J. Sanchez-Herencia, *Materials Letters*, 1992, 14, 5/6, 333.

14. W.E. Frazier and M.E. Donnellan, *Journal of Materials*, 1991, 43, 5, 10.

15. D.R. Clarke, *Journal of the American Ceramics Society*, 1992, 75, 4, 739.

16. M. Hunt, *Machine Design*, 1993, 26, 52.

17. W.D. Callister Jr., *Materials Science and Engineering*, 2nd Edition, J. Wiley and Sons Inc., New York, 1991.

18. R.P. Kambour, *Journal of Polymer Science-Reviews, Macromolecular Science*, 1974, 7, 1.

19. L.T. Manzione and J.K. Gillham, *Journal of Applied Polymer Science*, 1981, 26, 3, 889.

20. L.T. Manzione and J.K. Gillham, *Journal of Applied Polymer Science*, 1981, 26, 3, 907.

2 Constituent Materials

G. Akovali and C. Kaynak

2.1 Introduction

A composite, as described in the previous chapter, is anisotropic and is composed of at least two constituent materials that are properly bonded together: one constituent serving as a matrix surrounding the reinforcing particles or fibres of the other(s). Fibreglass reinforced (FGR) polyester is a common example where a tough polyester (matrix) is combined with high modulus-high strength (and brittle) glass fibres. This leads to a system with a collection of the most advantageous properties of individual constituents (i.e., better overall strengths and toughnesses) rather than an averages. Conversely, ACM are a highly specialised group of composites that are primarily identified by their aircraft/aerospace and leisure or sports applications. They can provide high performances under extreme mechanical, electrical and environmental conditions (which are projected for the year 2000 to be 30,000 tonnes, with a corresponding value of about 3–5 billion US Dollars, pointing at it's high value characteristics [1]). Today, the fraction of composites (mostly carbon fibre composites) in large passenger aircraft like Airbus A-320, Tupolev Tu-204 or Ilyushin Il-86 has reached up to 15% of the structure weight; the percentage by weight share of composites in military aircraft is already greater, and its use in military helicopters exceeded half of the total structure weight. ACM consist of a high strength reinforcing constituent combined with a high performance matrix, which is usually polymeric. Low weight and low thermal expansion, in addition to high stiffness, strength and fatigue resistances, are basic requirements which are mostly fulfilled by use of polymeric systems, giving them properties comparable with steel or aluminium. Chapter 2 will provide more detailed information on matrix polymers, as well as for reinforcing agents commonly used with their composites.

2.2 Matrix polymers

The matrix in a composite is the continuous phase providing uniform load distribution to the reinforcing constituent(s). The matrix, in addition to protecting the reinforcing constituent(s), safeguards the composite surface against abrasion, mechanical damage and environmental corrosion. Uniform load distribution is due to the existence of proper adhesion between dissimilar constituents, the extent of which may be required as minimal, i.e., for ballistic components, or maximal as in most cases, depending on the application. The matrix itself should provide enough fracture toughness or ductility and optimum hot/wet performance. Since the ultimate thermo-mechanical characteristics of the composite are principally governed by that of the matrix, heat resistance and thermal properties of the latter are also very important. In addition, a high performance matrix resin is expected to have a modulus of at least 3 GPa for strength and a sufficiently high shear modulus to prevent buckling of fibre-reinforcing constituents especially when under compression, although the matrix is known to play a minor role in the tensile load-carrying capacity of a composite. The matrix has a major influence on the interlaminar shear (particularly important for structures under bending loads), and on the in-plane shear properties (important for structures under torsional loads).

Table 2.1 Some advantages and disadvantages of using thermoset and thermoplastic matrices			
Application	**Property**	**Thermoset Matrix**	**Thermoplastic Matrix**
Matrix	Formulation	Complex	More Simple
	Melt Viscosity	Low (at the Beginning)	Rather High
	Fibre Impregnation	Comparably Easy	Very Difficult
	Cost	Low to Medium	Low to High
Prepreg	Tack/Drape	Good	Comparably Low
	Shelf Life	Very Poor	Good
Composite	Processing Cycle	Very Long	Short to Long
	Processing Temperature and Pressure	Moderate	High
	Size of Products	Can be Very Large	Small to Medium
	Resistance to Solvents	Good	Poor to Good
	Damage Tolerance	Poor to Excellent	Fair to Good
	Resistance to Creep	Good	Not Known
	Interlaminar Fracture		
	Toughness	Low	High
	Ease of Fabrication	Labour Intensive	Less Labour Intensive

Matrix polymers can be thermoset or thermoplastic in nature and they can also be rubbery. Quite a large variety of different matrix systems are available, each with certain advantages and disadvantages, as presented in **Table 2.1**. Matrix polymers as the raw material usually constitute some 40% of the total cost of the composite, followed by 30% for the fabrication costs [2].

There are a number of different properties expected to be fulfilled by the polymer matrix, depending on the application of the composite. To name a few, the matrix should:

a) wet and bond to the second constituent

b) flow easily for complete penetration and elimination of voids

c) be sufficiently elastic

d) have low shrinkage, low CTE

e) be easily processable and

f) have proper chemical resistance, low- and high-temperature capabilities, dimensional stability, etc.

2.2.1 Thermosets

Thermosetting polymers are the most frequently used matrix materials in polymer-based composites production, mainly because of the ease of their processing. They are low molecular weight telechelic reactive oligomers at the beginning. In general, they contain two (telechelic oligomer and curing agent) or more components, which consist of a proper catalyst and/or a hardener. Solidification begins when the components are mixed (at either ambient or elevated cure temperatures). The subsequent reaction produces a rigid, highly-crosslinked network or a vitrified system. During curing, there are various intermediate stages from liquid to gel and to the vitrified states, as characterised by time-temperature-transformation (TTT) isothermal cure diagrams [3]. From TTT diagrams and rheological or dynamic mechanical data, curing characteristics of thermosettings are optimised [4]. After setting at a specific temperature (ambient or above), thermosets cannot be reshaped by subsequent heating.

Thermoset matrix polymers are usually considered in three groups:

a) low temperature thermoset matrices (mainly polyesters),

b) medium temperature thermoset matrices (mainly phenolics) and

c) high temperature thermoset matrices (polyimide and bis-maleimides) [7]. Within these, the most commonly used thermoset polymers (as composite matrices), are epoxides and polyesters, followed by polyimides and bismaleimides. Epoxides are briefly described in section 2.2.1.1.

2.2.1.1 Epoxy

The starting materials for the epoxy matrix are low molecular weight organic liquid resins containing a number of epoxide groups, which are simply three-membered rings with one oxygen and two carbon atoms (**Figure 2.1**).

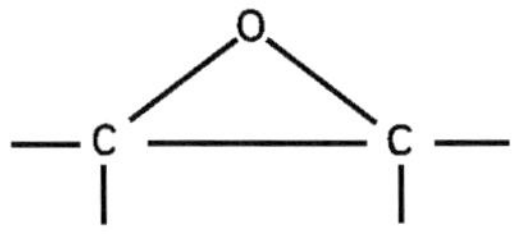

Figure 2.1 Characteristic group for epoxies

Epoxides can vary from difunctional to polyfunctional and, after their reaction with the curing agents (which occurs without the evolution of any by-product), they yield to high performance systems with a combination of high rigidity, solvent resistance and elevated temperature behaviour. The most widely used difunctional epoxy is the diglycidyl ether of bisphenol A, with (n) from 0.2 to 12 (formula 1 in **Figure 2.2**) with use of different

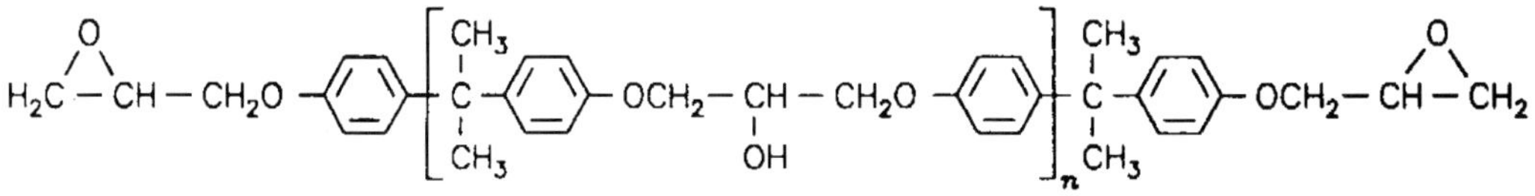

(1)

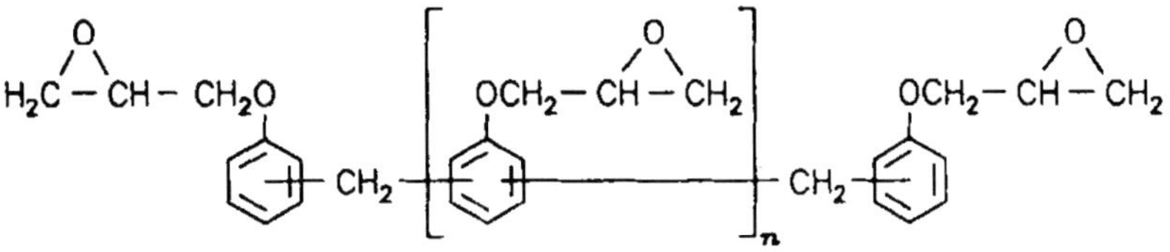

(2)

Figure 2.2 Difunctional epoxy (1) and epoxidised novolak (2)

types of curing agents, i.e., various amines. Epoxidised novolaks (formula 2), possess multi-epoxy functionality of at least 2, or more than 5, epoxy groups per molecule.

For high temperature aerospace applications, a special polyfunctional epoxy with aromatic and heterocyclic glycidyl amine groups, i.e., tetraglycidyl methylene dianiline, (TGMDA; formula 3 in **Figure 2.3**) is used.

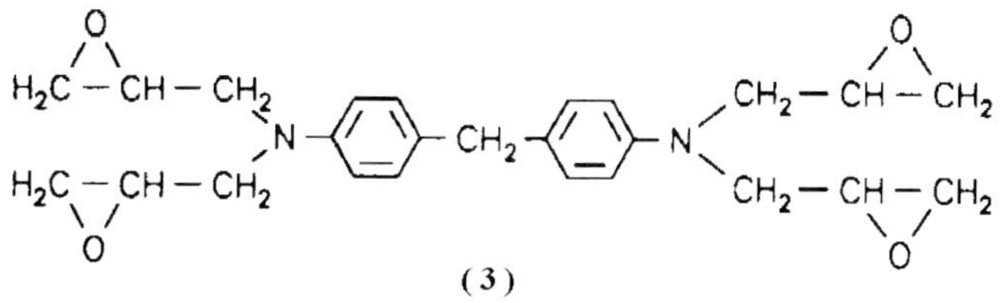

Figure 2.3 Tetraglycidyl methylene dianiline (TGMDA)

A new generation of aromatic and glycidyl amine resins with improved hot/wet temperature characteristics are also available (formulae 4 and 5 in **Figure 2.4**). There are, in addition, a number of special multi-functional epoxides formulated with TGMDA and/or bisphenol A [1].

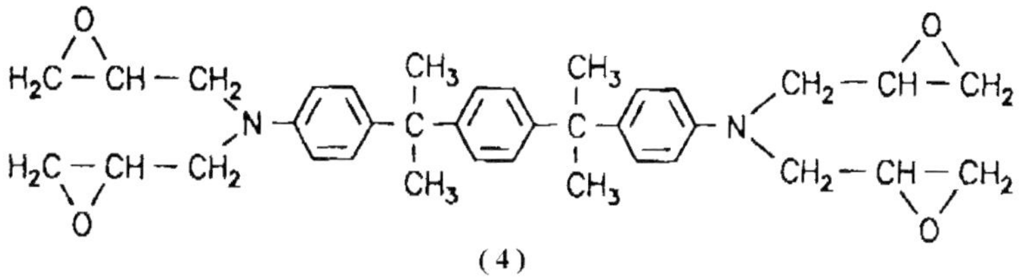

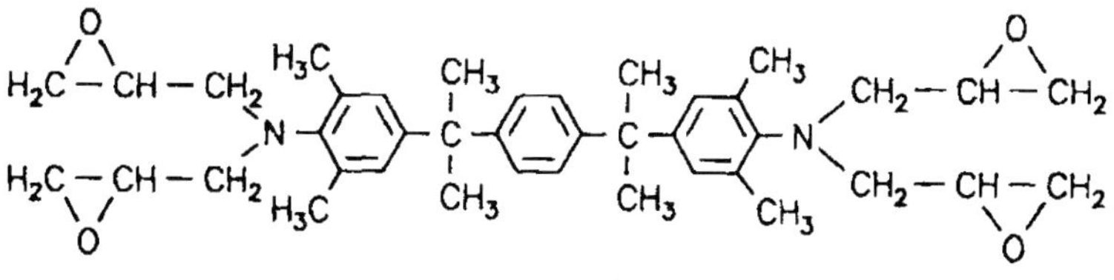

Figure 2.4 New generation aromatic and glycidyl amine resins

Curing agents used for epoxides are either co-reactants (become incorporated with the epoxide during the reaction), or act as a catalyst to promote crosslinking. The first type of curing agents are polyfunctional and can be basic (primary/secondary amines, polyaminoamides) or acidic (anhydrides, polyphenols, polymeric thiols). Depending on the basicity or acidity of curing agent, curing may occur at room or at high temperatures. Catalytic curing agents (tertiary amines and BF_3 complexes) can accelerate curing at low or at ambient temperatures.

Selection of the epoxy resin for any composite application is usually done by considering the final application conditions of the composite since there are significant differences existing between thermal and mechanical properties of different epoxides, i.e., in moduli, in strain to failure and in T_g, and usually there is a compromise between the high temperatures and toughnesses. That is, the T_g controls the application temperature, which is high (247 °C) for brittle epoxides, but much lower for toughened epoxides. The degree of polymerisation also can effect processability and crosslink densities [8]. In addition, the type of curing agent or accelerator and its molar ratio to the epoxy affects some of the final properties of the epoxy system, and these must be optimised, because of a change in crosslink densities. For structural applications, the hardener dicyandiamide (DICY), is usually used. By careful selection of the polymer and curing agent to accelerator ratio, and their use with the optimised values, the properties of the final product can be assured. In addition, there are a number of methods and strategies which have been developed for combination of the most desirable features of a thermoplastic resin segment within the epoxy resin to form a multi-component composite system [5]. A considerable amount of research has been done to further improve the toughness [9, 11], moisture resistance [10] and heat stability of epoxy matrices. These studies led to the production of epoxy composite systems for subsonic aircrafts with the desired tensile-compressive moduli and tensile strength values of 138 GPa and 1930 MPa [12]. In comparison, a tensile-compression modulus of 3100 MPa and a tensile strength value of 96 MPa was recorded for the neat resin.

Epoxy resins have the added advantage over many other thermosets in that, since no volatiles and condensation products other than the polymer product are produced during cure, moulding does not require, in principle, high pressure moulding equipment.

By use of the epoxy matrix, one can gain a system with following advantages:

a) a wide variety of properties,

b) low shrinkage during cure (lowest within thermosets),

c) good resistance to most chemicals,

d) good adhesion to most fibre, fillers,

e) good resistance to creep and fatigue, and

f) good electrical properties.

Conversely, the following principal disadvantages may apply:

a) sensitivity to moisture (after moisture absorption (1%–6%), there usually is a decrease in the following: heat distortion point, dimensions and physical properties,

b) difficulty in combining toughness and high-temperature resistances (as explained above),

c) high CTE as compared to other thermosets,

d) susceptibility to UV degradation, and

e) cost (epoxies are more expensive than polyesters) [13].

2.2.1.2 Polyester (unsaturated)

Unsaturated polyesters are the most versatile class of thermosetting polymers. They are macromolecules consisting of an unsaturated component, (i.e., maleic anhydride or its trans isomer, fumaric acid; which provides the sites for further reaction), and a saturated dibasic acid or anhydride with dihydric alcohols or oxides (typically phthalic anhydride, which can be replaced by an aliphatic acid, like adipic acid, for improved flexibility). If blends of phthalic anhydride and maleic anhydride/fumaric acid are used, 'ortho resins' (**Figure 2.5**) are produced. If isophthalic acid and maleic anhydride/fumaric acid are

Prepolymer Polyester

+

Styrene + Catalyst (+ Heat)

Crosslinked Network

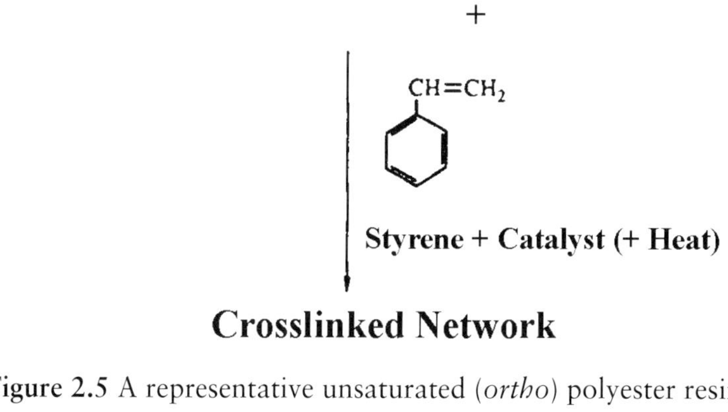

Figure 2.5 A representative unsaturated (*ortho*) polyester resin

27

used, 'iso resins' result. Conversely, if propoxylated or ethoxylated bisphenol A is used with fumaric acid, 'bisphenol A fumarates' are produced. Also, if a blend of chlorendic anhydride or chlorendic acid and 'maleic anhydride/fumaric acid are used, 'chlorendics' result. Within these, biphenol A fumarates are unique for high performance and chlorendics for improved flame-resistance applications. If the unsaturated resin is prepared by the reaction of a monofunctional unsaturated acid (i.e., methacrylic or acrylic acid) with a biphenol diepoxide, a new family of polyesters with exceptional mechanical and thermal properties is obtained, and they are termed 'vinyl resins' [7].

In most cases, the polymer (polyester), is dissolved in a reactive vinyl monomer, i.e., styrene, to give a proper solution viscosity. The resin is cured by use of a free radical catalyst, the decomposition rate of which determines the curing time. Hence, curing time can be decreased by increasing the temperature (for a high temperature cure, say at 100 °C, benzoyl peroxide is commonly used; whereas for a room temperature cure, other peroxides with metal salt accelerators are preferred). Crosslinking reactions occur between the unsaturated polymer and the unsaturated monomer, converting the low viscosity solution into a three-dimensional network system. Crosslink densities can change (by direct proportionality), the modulus, T_g and thermal stabilities; and (by inverse proportionality), strain to failure and impact energies. The formation of the crosslinked structure is accompanied by some volume contraction (7%–27%).

Polyesters, if used as a matrix for composite production, have improved tensile and flexural strength values and after extensive annealing treatment, their sensitivity to brittle

Table 2.2 Some characteristics and uses of epoxy and polyester thermosets			
Thermoset	Some Characteristics	Main Uses	Limitations
Epoxy	Good Electrical Properties Chemical Resistance High Strength	Filament Winding Printed Circuit-Board Tooling	Require Heat Curing for Maximum Performance Cost
Polyester	Good All-Around Properties Ease of Fabrication Low Cost Versatile	Corrugated Sheeting Boats, Piping, Tanks	Ease of Degradation

fracture is usually improved and even eliminated [6]. Because of the high content of aromatic vinyl groups, the crosslinked polyester is easily susceptible to thermooxidative decomposition, which reduces the long-term application temperature. In general, they have good chemical and corrosion resistance, as well as good outdoor resistance. As in the case of epoxies, polyesters have the added advantage over many other thermosets in that they do not require high pressure moulding equipment. **Table 2.2** presents some characteristics and uses of polyester and epoxy thermosets.

2.2.1.3 Polyimides and bismaleimides

The polyimide matrices are mostly used in premium high performance composite applications as they are speciality polymers. Their tensile and flexural strengths are commonly around 110 and 200 MPa, respectively. Polyimides are (a) condensation polyimides, (b) addition polyimides or (c) thermoplastic polyimides. The first and third of these are thermosets while the second is a thermoplastic.

Condensation polyimides are condensation products of difunctional carboxylic acid dianhydrides and primary diamines, and contain (—CO—NR—CO—) groups as a linear or heterocyclic unit along the polymer backbone. They may be translucent or opaque. A condensation polyimide is shown in **Figure 2.6**.

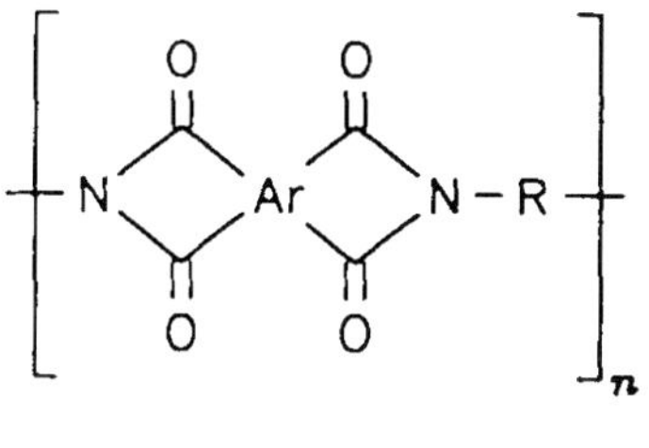

Figure 2.6 A condensation polyimide (Ar denotes aromatic)

Mechanical properties and thermooxidative stabilities for the aromatic, heterocyclic polyimides (**Figure 2.7**) are outstanding, and they are used for some high performance applications in place of metals and glass [24]. However, their prices are rather high.

Bismaleimides (**Figure 2.8**) can be considered as structurally-modified polyimides or as polymerisation of monomer reactants (PMR) [6]. They are also known as addition

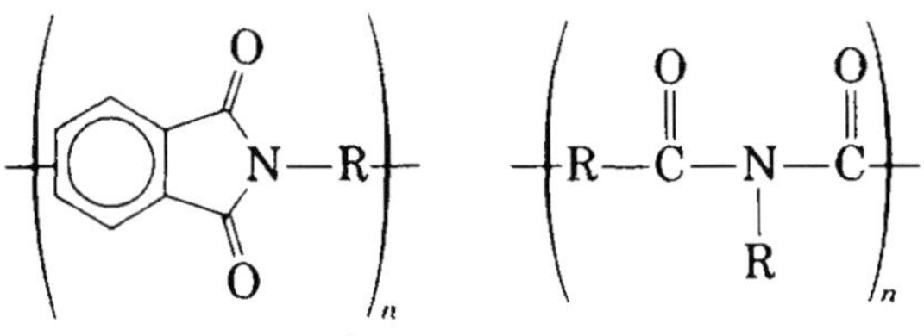

Figure 2.7 An aromatic, heterocyclic polyimide

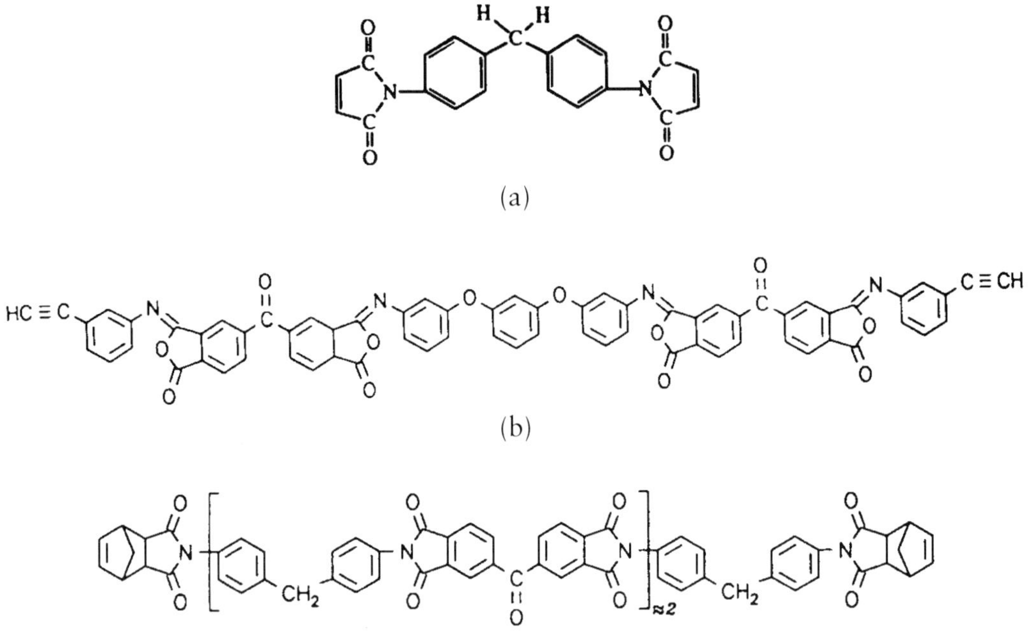

(a)

(b)

(c)

Figure.2.8 (a) Bismaleimide, (b) Thermid, and (c) Norbornene

polyimides, which are obtained by using end-capped imide oligomers with unsaturated functional groups like olefins, norborene, acetylene or maleimide. These are called thermides (**Figure 2.8b**), norborenes (**Figure 2.8c**) and bismalemides respectively, and are curable by an addition mechanism. The thermally-induced crosslinking reaction occurs via a free radical mechanism across the terminal double bonds, which can react with themselves or with other co-reactants (such as vinyl, allyl or functionalities). PMR have developed as a result of the demands of certain industries, particularly the aerospace sector. They are materials with low flammability, high strength and high mechanical and thermal integrity at high temperatures in aggressive environments for prolonged periods of time. However, PMR possess inherent brittleness, which is modified by the use of several thermoplastics to increase their toughness.

2.2.2 Thermoplastics

Thermoplastics are heat softenable, heat meltable and reprocessable. It is generally recognised that the use of thermoplastics in composite production results in a decrease in manufacturing costs. Other factors include;

a) the indefinite prepreg stability without refrigeration,

b) a relatively fast processing cycle,

c) easy quality control and quality assurance, and

d) the reprosessability of the moulded compound to correct imperfections.

Usually, there are higher levels of toughness values and proper damage tolerances for thermoplastics as compared with thermosets. However, being partially crystalline for most cases, the resulting polymer matrix may also suffer from a change in the level of crystallinities during the fabrication and processing of composites due to variations in the heating and cooling rates, as well as the thickness. These factors may affect the polymer's mechanical and physical properties considerably. Since the emphasis is for the development of tougher composite materials with reduced weight, for low costs and higher damage tolerances, most thermoplastics meet these demands. Within thermoplastic matrix materials, a number of different thermoplastics with demanding performances can be cited; such as thermoplastic polyesters, polyamides, polysulphones, polyaryl ethers, thermoplastic polyimides, polyarylene sulphide and liquid crystalline polymers.

2.2.2.1 Thermoplastic polyesters (TPE)

Polyesters are based on phthalates, and contain the ester group (—COO—) in the main chain. There are two major groups of polyesters in the TPE family: aromatic-aliphatic and aromatic. Thus classification is based on the type of copolymerised dial. Poly(ethylene terephthalate) (PET), and poly(butylene terephthalate) (PBT), are both members of the aromatic-aliphatic polyester family.

HO CH$_2$ CH$_2$ O (OC - C$_6$H$_4$ - COO CH$_2$ CH$_2$ O)$_n$H   (PET)

HO CH$_2$ CH$_2$ O (OC - C$_6$H$_4$ - COO CH$_2$ CH$_2$ CH$_2$ CH$_2$ O)$_n$H   (PBT)

PET is hard, rigid and exhibits very little wear and tear. It also has very little creep and can tolerate very high mechanical loads. Polybutylene terephthalate (PBT) and can be processed at lower temperatures than PET, but its T$_g$ is lower and has somewhat poorer mechanical properties. The service temperature of neat polyester matrix ranges from 120 to 240 °C. Because of their high crystalline melting points and T$_g$, they retain their good mechanical properties at high temperatures. Their chemical and solvent resistances are good.

Aromatic polyesters belong to the second group of TPE polyesters, and are also known as polyarylates (**Figure 2.9**). They are produced by the combination of bisphenol A with isophthalic or terephthalic acid. Polyarylates are flame-retardant and have good mechanical and electrical properties. However, they are sensitive to heat, and although their mechanical properties are not affected significantly by heat, their colours darken.

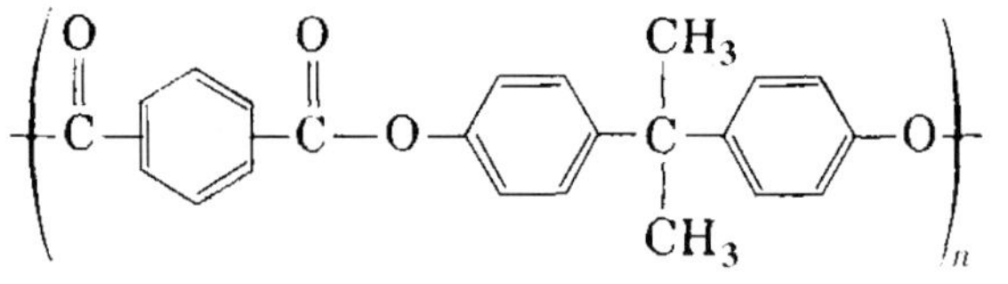

Figure 2.9 A polyarylate

Another member of the TPE family that should be mentioned is Xydar (Dart Industries), which is a commercial liquid crystal polymer (LCP) with a higher heat deflection temperature (355 °C for the latter as compared with 55 °C for PET) and processing melt temperature (450 °C against 270 °C for PET). The monomeric units that comprise Xydar are p-phenyl bisphenol (PPPB), p-hydroxy benzoic acid (HBA) and terephthalic acid (TPA). The structure of Xydar is shown as formula 1 in **Figure 2.10**. A similar thermotropic LCP is Vectra (from Celanese Industries), which is based on hydroxy naphthanoic acid (HNA), TPA and HBA (formula 2 in **Figure 2.10**).

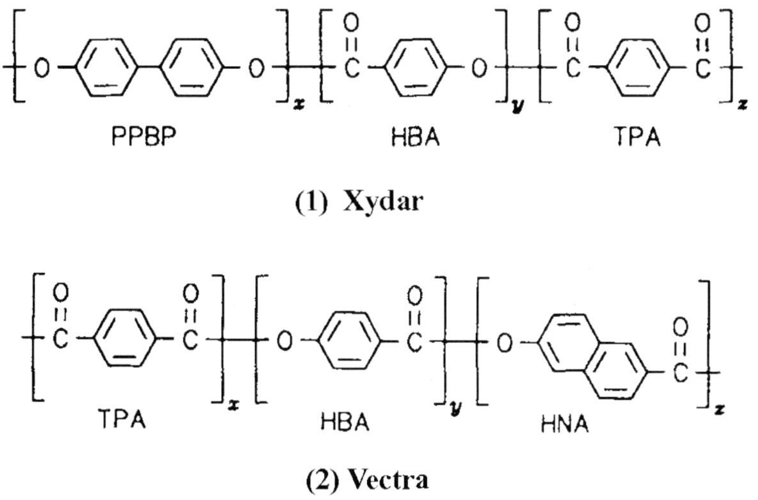

(1) Xydar

(2) Vectra

Figure 2.10 Xydar and Vectra

2.2.2.2 Polyamides (PA)

The word Nylon has been accepted as a generic name for synthetic polyamides, and they are described by a numbering system indicating the number of carbon atoms in the monomer chains. Amino acid polymers are designated by a single number, as Nylon-6 for poly(ω-aminocaproic acid) or polycaprolactam. Nylons from diamines and dibasic acids are designated by two numbers, the first representing the diamine while the second is the acid involved (as Nylon-66 for the polymer of hexamethylene-diamine and adipic acid, Nylon-610 for hexamethylene-diamine and sebacic acid, respectively).

In general, polyamides have a combination of high strength, elasticity, toughness and abrasion resistance. Mechanical properties are usually maintained up to 150 °C and its toughness and flexibilities are retained well at low temperatures. Higher aliphatic Nylons (like 610, 612, 11 and 12) show lower stiffness and heat resistances than lower Nylons (6 and 66) (**Figure 2.11**), but they have improved chemical resistance and much lower moisture sensitivities.

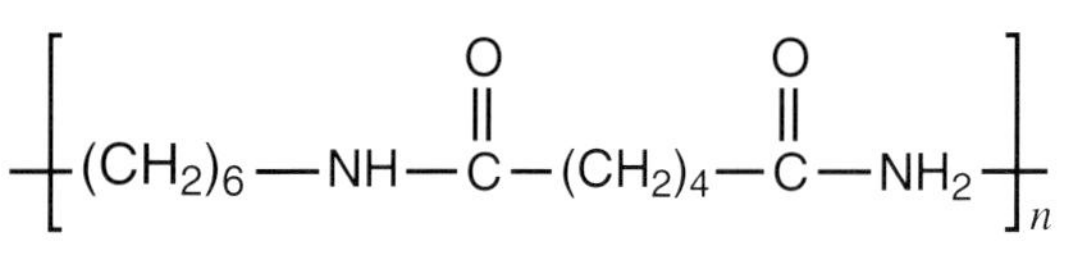

Figure 2.11 Nylon-66

To tailor the properties of polyamides for specific applications, their blends and alloys with polyolefins and certain copolymers and epoxies [21, 22] are also prepared and used.

In addition to the group of aliphatic polyamides, there is also a high strength aromatic polyamide family, known generically as aramids. One example is poly-metaphenylene (MPD), which has a trade name of Nomex and is in **Figure 2.12**. Similarly, there is the *p*-phenylene phthalamide (PPD), which has a trade name of Kevlar (Figure 2.12). Nomex and Kevlar are widely used in composite production, with the former being virtually non-flammable and Kevlar possessing an outstanding strength to weight ratio. Kevlar cable can match the strength of steel at the same diameter, yet its weight is 20% lower. The *para*-aramid molecule is rigid due to the para substituents of the aryl rings in addition to the C—N bond, which limits rotation and hydrogen bonding between amide groups, all of which contribute to the stiffness.

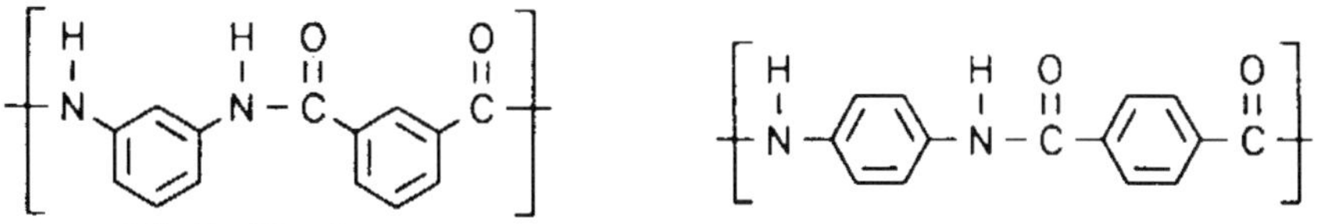

MPD (Kevlar)   **PPD (Nomex)**

Figure 2.12 Kevlar and Nomex

2.2.2.3 Polyaryl ethers

In this group, the most common examples are polysulphone (PS) and PEEK.

2.2.2.3.1 PS

PS are a group of amorphous aromatic thermoplastics with T_g of around 185 °C and continuous application temperatures of 160 °C. PS have good resistance to most mineral acids, alkalis and salt solutions, however, they swell, stress crack or dissolve in polar organics. The general formula of PS is where R and R′ contain aromatic rings and ether linkages (**Figure 2.13**).

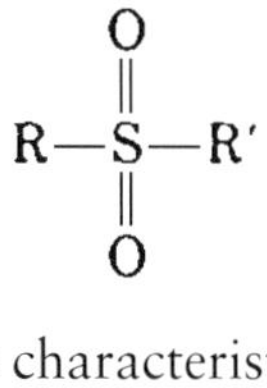

Figure 2.13 The characteristic group for PS

In the PS family, there are polyaryl sulphones and polyaryl ether sulphones, which exhibit outstanding heat resistance, high resistance to creep, rigidity and transparency (**Figure 2.14**).

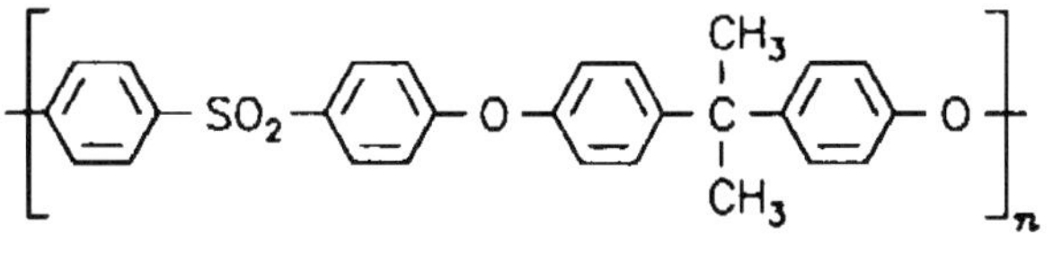

Figure 2.14 Polysulphone molecule with diphenylene sulphone group

2.2.2.3.2 PEEK

PEEK (**Figure 2.15**) is a crystalline grey resin with outstanding solvent resistance and high melting point (290 °C), high thermal stability (maintains ductility if kept at 200 °C for years), high hydrolytic stability and with flame-retardant properties. PEEK is used predominantly for aerospace and military applications [23].

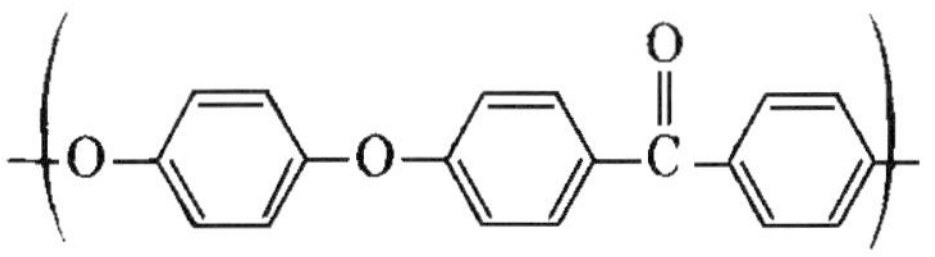

Figure 2.15 PEEK

2.2.2.4 Thermoplastic polyimides

Thermoplastic polyimides are linear polymers derived by polymerisation of a polyamic acid and an alcohol and, depending on the types of both reacting components, different polyimides are produced. They are used when heat and environmental resistances are required. The most common type of thermoplastic polyimides are polyether imide (PEI), polyamide imide (PAI), K polymers and LARC (Langley Research Center thermoplastic imide); the latter two are usually available as prepolymers. The open formulae of PEI and PAI are given in **Figure 2.16a and 2.16b.**

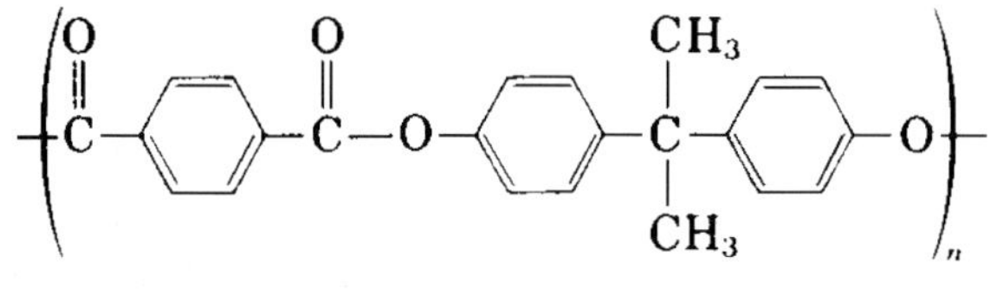

Figure 2.16a Open formula of polyether imide (PEI)

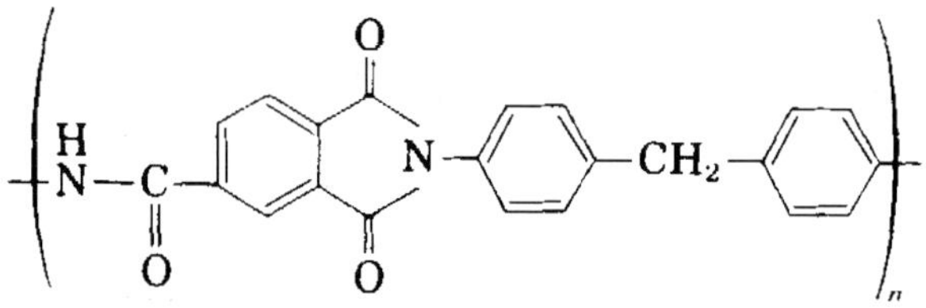

Figure 2.16b Open formula of polyamide imide (PAI)

2.2.2.5 Polyaryl sulphides

PPS is a member of the polyaryl sulphide family. PPS (**Figure 2.17**) is a semi-crystalline polymer with a melting point of 287 °C. The continuous application temperature of the resin is 240 °C. PPS can crosslink with air and oxidise by itself at elevated temperatures. It has outstanding resistance to heat and chemicals, with excellent electrical insulation characteristics. It tends to be brittle, and its mechanical properties, as well as its mould shinkage, can be improved by application of fibre reinforcement. PPS is one of the most expensive thermoplastic polymers.

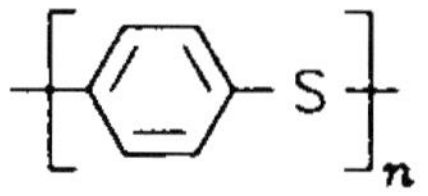

Figure 2.17 PPS

See Chapter 1 for some typical properties of certain thermoplastic matrix materials.

2.3 Reinforcing agents

There are quite a large number of different substances that are used as 'fillers' for plastics, mostly for 'cost reduction' purposes. Reinforcing agents are a special class of fillers with 'reinforcement' properties of the matrix, and they are either particulates or fibrous in shape, with classification depending mainly on the aspect ratios (ratio of length to breadth) involved.

2.3.1 Particulate reinforcing agents

Particulate reinforcing agents have similar length and breadths with aspect ratios around '1'. They cover particulates of regular shape, (such as spheres), as well as those with irregular shapes that may have extensive convolution and porosity. Particulate reinforcing agents are the most common and the cheapest materials. They can produce the most isotropic property of composites. Particulate composite reinforcing agents can be divided into two groups: organic and inorganic. Organic particulate reinforcing agents usually have limited thermal stability with low densities. Some of the important examples to mention are: powdered cellulose, powdered rubber, a wide variety of starches and particulate carbon [14]. Silica and some simple metal oxides (like alumina), glass and even clay are examples of inorganic reinforcing particulates with the feature that they reinforce the matrix. Some additional information on particulate reinforcing agents was provided in Chapter 1.

2.3.2 Fibre reinforcing agents

Fibre reinforcing agents can be either organic or inorganic. Of the natural (organic) reinforcing fibres, cellulosics are the most important applied as chopped cloth, fabric or yarn. The ligno-cellulosic fibres like jute and sisal are widely used in phenolic and polyester systems. Within the organic synthetic fibres, one can mention: nylons (specifically in phenolic ablative systems), high- and ultra-high modulus polyethylene (with moduli in the range of 50–100 GPa, about ten times those of normal textile fibres), polypropylene (in inorganic cement) and aramids (Kevlar and Nomex); all applied as woven, non-woven fabrics, fibres or rovings. Carbon fibres are used to reinforce metals as well as polymeric matrices, and they are also considered as organic synthetic reinforcing agents. Examples of inorganic natural fibres are asbestos (which is no longer used due to the health and safety issues) and glass fibres. More specific examples of inorganic fibres are boron and ceramic fibres, in particular those based on alumina. In addition, there are various metal filaments, i.e., aluminium, copper, steel, etc., in the form of fibres, wires and whiskers.

Reinforcement of metallic, polymeric or ceramic matrices by fibres have certain differences. In the case of polymer matrices, the organic phase is expected to coat the fibre surfaces, and the interactions at the interface are physical in character. Conversely, in ceramic composites, a rather weak interaction at the interface is desirable to achieve the improved fracture toughnesses. In metallic matrices, however, a very strong interaction and bond at the atomic level at the interface is essential, and therefore a high degree of compatibility at the interface is needed.

2.3.2.1 Glass fibres

The use of glass fibres dates back to the ancient Egyptians [15]. Glass is an amorphous material composed of a silica network. There are four main classes of glass used commercially: high alkali (essentially soda-lime-silica: A glass), electrical grade (a calcium alumino-borosilicate with low alkali oxide content: E glass), chemically-resistant modified E glass grade (with calcium alumino-silicate: ECR glass) and high strength grade (with magnesium alumino-silicate and no boron oxide: S glass). Fibres from any of these can be prepared, however, E glass fibre is the one most widely-used for reinforcement purposes, although S glass fibre has the highest tensile strength and elastic modulus of these four types of glasses (**Table 2.3**). Glass fibre is spun from the melt and it is obtained after cooling it to the rigid condition without crystallising. Once the continuous glass fibres are produced, they are transformed into one of the finished forms, which are (continuous or woven) rovings, yarns (mostly for textile applications), chopped strands, (fibreglass) mats and preforms.

Table 2.3 Typical properties of different glass fibres					
Material	Density (kg/m^3)	Tensile Strength (MPa)	Young's Modulus (GPa)	CTE (10^{-6}/K)	Strain to Failure (%)
E-Glass	2620	3450	81	5.0	4.9
S-Glass	2500	4590	89	5.6	5.7
A-Glass	2500	3050	69	8.6	5.0

In all fibrous composite technologies, a proper fibre sizing (fibre finishing or application of a coupling agent) is essential; which is any surface coating applied to a reinforcement to protect it from damage during processing, to aid in processing and to promote adhesion to the matrix. They can be film-forming organics and polymers, adhesion promoters (like silane coupling agents) or chemical modifiers (like silicone carbide on boron fibres). The latter process must not be considered simply as a surface treatment, since there is always a chemical modification and altering of surface chemical composition, with or without a coating at the surface [16].

2.3.2.2 Carbon fibres

Carbon fibres have been known for more than 100 years, however it was only after the 1950s that the increase of interest for high strength and lightweight reinforcements in aerospace applications made their use more common. They are obtained by the pyrolysis of organic precursor fibres such as rayon, polyacrylonitrile (PAN), or pitch. Carbon fibres are typically carbonised between 1200–1400 °C and contain 92%–95% carbon regularly in the non-graphitic stage; that is, with two-dimensional long range order of the carbon atoms in planar hexagonal networks, but without any measurable crystallographic order in the third direction [17]. After carbonisation, tensile strength values of 3000 MPa and moduli of 250 GPa or even higher are usually achieved, the latter of which can even be improved up to 350 GPa, at the expense of some drop in strength by the post treatment.

The term carbon fibre is often used interchangeably with the term graphite fibre, although they differ from each other. Graphite fibre is produced by graphitising the precursors at a much higher temperature (1900 to 2500 °C) and contains more elemental carbon (more than 99%). Carbon fibres are used as yarn, felt or powder-like short monofilaments with diameters smaller than 10 mm. There are different types of carbon fibres depending on the origin of precursor, such as (a) PAN-based, (b) isotropic pitch-based, (c) anisotropic pitch-based, (d) rayon-based, or (e) gas phase grown [17]. Depending on mechanical properties, carbon fibres are classified into four categories: (a) PAN-based high modulus (HM)-low strain to failure- type (b) PAN-based high tensile (HT)-high strain to failure-type and (c) PAN-based intermediate modulus (IM) type and (d) mesophase (pitch)-based. The third of these belong to the HT type with high tensile strength and improved stiffness. **Table 2.4** presents some mechanical properties of different types of carbon fibres.

Table 2.4 Tensile properties of carbon fibres			
Fibre Type	Young's Modulus (GPa)	Tensile Strength (GPa)	Strain to Failure (%)
PAN-based High Modulus	350–550	1.9–3.7	0.4–0.7
PAN-based Intermediate Modulus	230–300	3.1–4.4	1.3–1.6
PAN-based High Strength	240–300	4.3–7.1	1.7–2.4

2.3.2.3 Aramid/Kevlar fibres

Aramids are aromatic polyamides and the most common two members of this family are Kevlar and Nomex, which were first introduced in the early 1970s. In Kevlar (for formula, see Figure 2.12), the aromatic ring structure contributes to high thermal stability and the para configuration leads to rigid molecules that contribute high strength and modulus. They are characteristically liquid crystalline.

When a solution of Kevlar is extruded into fibre form, a highly anisotropic structure with an exceptional degree of alignment of straight polymer chains parallel to fibre axis develops, giving rise to higher strength and modulus in the fibre longitudinal direction. In addition, there is fibrillation in the structure, which has a very strong effect on fibre properties and failure mechanisms. Para-aramid fibres have good toughness and damage-tolerance characteristics. They do not have a conventional melting point or T_g (estimated as >375 °C) and decompose in air around 425 °C. They are flame-resistant. *Para*-aramid fibres have a small, but negative, longitudinal (and a bigger positive transverse) CTE value. The fibre can be degraded chemically only by strong acids and bases, in addition to ultraviolet radiation.

Within the different types of Kevlars, there is Kevlar 29 (with high toughness), Kevlar 149 (with ultra high modulus) and Kevlar 49 (with high modulus). In structural composite production, Kevlar 49 is the most dominant form used today. Each of these Kevlar fibres are also available in a range of different short fibre forms and yarn counts. For Kevlar fibres, moduli range between 83–186 GPa, while tensile strengths are around 3.4 GPa, the latter of which is more than twice the strength of Nylon-66 and 50% greater than that of E-glass. **Table 2.5** presents some characteristics of different *para*-aramid fibres.

Table 2.5 Some properties of *para*-aramid fibres				
Fibre Type	Density (kg/cm³)	Young's Modulus (GPa)	Tensile Strength (GPa)	Strain to Failure (%)
Kevlar 29 (High Toughness)	1440	85	3.0–3.6	4.0
Kevlar 49 (High Modulus)	1440	131	3.6–4.1	2.8
Kevlar 149 (Ultra High Modulus)	1470	186	3.5	2.0

2.3.2.4 Boron fibres

Boron monofilament continuous fibres are produced by the chemical vapour deposition (CVD) method, and are mostly used to reinforce epoxy matrices (to form a boron-epoxy prepreg tape), for sporting goods as well as in aerospace applications. They were specifically used in the space shuttle. The structure of the fibre depends on the deposition conditions, especially temperature. At deposition temperatures below 1300 °C, the fibres are amorphous; at higher temperatures crystalline boron fibre is obtained. The fibre has high strength, high stiffness and low density with a 'corncob structure' surface appearance, consisting of modules separated by boundaries. If fibre surfaces are smoothed, i.e., by chemical etching, the average tensile strengths can be doubled [18].

2.3.2.5 Other polymeric fibres

Natural polymeric fibres, mostly cellulosics, such as flax, jute and cotton, have been used since ancient times, some key mechanical properties of which are inferior to glass, carbon or aramid fibres. Flocks of cotton, as well as chopped cloth and yarn, of cellulosics are the most common. In the woven state, cellulosics are often used as a laminating material. After availability of synthetics, a number of different types of reinforcing fibres from Nylons, acrylics, polyolefins, etc., are produced by spinning. For instance, after the introduction of ultra-high modulus fibres (in particular from polyethylene), a new generation of synthetic organic fibres with moduli in the range of 50–100 GPa (approximately ten times those of conventional fibres) were achieved. Although Nylon fibres have found application in phenolic ablative plastics, and polypropylene fibres are used to reinforce cement, the extent of reinforcements are limited. Ultra-high modulus fibres have opened new opportunities in that sense. The recent introduction of PEI fibres by Akzo (Enka) in filament and staple forms, offer high temperature and good environmental resistances, while para-phenylene polybenzobisoxazole (PBO) fibres by Dow have a unique combination of high strength, stiffness and environmental resistance. PBO offers tensile strength and modulus values that are higher than Kevlar. Another example is the polybenzazoles (PBZ) family, which have good strength and moduli values.

2.3.2.6 Ceramic fibres

Ceramic fibres are polycrystalline refractory materials composed of various metal oxides, metal carbides, metal nitrides and their mixtures. The first group of ceramic fibres are based upon alumina with polycrystalline structures. They are known to have better transverse properties than *para*-aramid or carbon fibres, which is due to their polycrystalline structure. Ceramic fibres are available in different fibre lengths and as fabrics. Continuous aluminium oxide fibres, commercial fibre FP (specifically of dense alumina fibres with the alpha crystalline form), has high modulus of elasticity, high melting point and good resistance to corrosives. Alumina silicate ceramic fibres with significant quantities of boria as continuous filament yarns are produced in braided and woven forms. They are the most advanced high performance (all crystalline) fibres available commercially and are known as Sumitomo and Nextel ceramic fibres. Nextel ceramic fibre contains considerable amounts of zirconium oxide and is called zirconia-silica ceramic fibre.

Other ceramic fibres include silicone nitride, silicone carbide, boron nitride, thoria, aluminium nitride, potassium titanate, high silica, quartz and a recently introduced polymeric material, polycarbosilane. This group of ceramic fibres are non-oxide fibres with much better mechanical properties than oxides (especially with higher elastic moduli and axial compression in the case of large diameter fibres), and both are prepared by CVD. Silicone carbide fibres have potentially low cost and are used mostly for ceramic matrix composites and metal matrix composites, and are commercially available as continuous and discontinuous (whisker). The term whisker applies to single crystals with fibrous or fibrillar characteristics. They have higher tensile strength and moduli as compared with polycrystalline continuous or discontinuous counterparts.

The densities of ceramic fibres are higher than glass and carbon fibres.

2.3.2.7 Metallic fibres

Metallic reinforcing fibres can be in the form of whiskers, metal wool and filament length fibres. Metals in wire or filament form exhibit considerable high strength, which is characterised by their high elastic moduli (i.e., beryllium, molybdenum, steel and tungsten). Because of its refractory properties, tungsten wire is used in some nickel- and cobalt-based superalloys, while steel wire is commonly used for concrete and tyres. The main disadvantage of wires (with the possible exception of beryllium), is their high densities. Since wires are, in general, more ductile than all other available fibres, it is of growing interest to produce composites that will carry high tensile loads with considerable toughnesses [19, 20].

2.4 Fibre forms

Fibres are combined with the polymer matrix in several forms, depending on the properties desired in the material and the processing method to be used to shape the composite. In some fabrication processes, fibres are continuous, while in others they are chopped into short lengths. In the continuous form, individual filaments are usually available as roving. A roving is a collection of untwisted (parallel) continuous strands; this is a convenient form for handling and processing. Rovings typically contain from 12 to 120 individual strands. By contrast, a yarn is a twisted collection of filaments. Continuous rovings are used in several polymeric composite processes, including filament winding and pultrusion.

The most familiar forms of continuous fibres are woven rovings and woven yarns, i.e., fabrics of rovings and yarns with various weave patterns. As a variation of these woven forms, rovings and yarns can be interwoven to produce braids, while knits can be obtained by interloping chains of rovings and yarns [25].

Fibres can also be prepared in the form of a mat: a felt consisting of randomly-oriented short fibres held together loosely with a binder, sometimes in a carrier fabric. Mats are commercially available as blankets of various weights, thickness, and widths. Mats can be cut and shaped for use as preforms in some closed-mould processes.

2.4.1 Rovings

Rovings (also called tows), refers to a group of essentially parallel strands of fibres that have been gathered into a ribbon and wound onto a cylindrical tube. This is called a multi-end roving process. The process begins by placing a number of oven-dried forming packages into a creel. The ends are then gathered together under tension and collected on a precision roving winder that has a constant traverse-to-winding ratio (the length of the strands traversed in one turn of the winder). This ratio has a significant effect on package stability, strand characteristics, and ease of subsequent operations [26].

Rovings are used in many applications. When used in a spray-up fabrication process, the roving is chopped with an air-powered gun which propels the chopped-glass strands into a mould, while simultaneously applying resin and catalyst in the correct ratio. This process is commonly-used for bathtubs, shower stalls, and many marine applications. In another important process, the production of sheet moulding compound (SMC), the roving is chopped onto a bed of polyester resin and compacted into a sheet. This sheet is then placed in a press and moulded into parts. Many fibre-reinforced plastic automotive body panels are made by this process.

Filament winding and pultrusion are processes that use rovings in continuous form. Applications include pipes, tanks, leaf springs (springs made of composite strips), and many other structural composites. In these processes, the roving is passed through a liquid resin bath and then shaped into a part by winding the resin-impregnated roving onto a mandrel or by pulling it through a heated die.

2.4.2 Yarns

Yarns are obtained by combining single strands by twisting and plying, i.e., they are produced by simply twisting two or more single strands together and subsequently plying. Plying essentially involves re-twisting the twisted strands in the opposite direction from the original twist [27].

Fine-fibre strands of yarn from the forming operation are air dried on the forming tubes to provide sufficient integrity to undergo a twisting operation. Twisting provides additional integrity to yarn before it is subjected to the weaving process, a typical twist consists of up to two turns per 5 cm. In many instances heavier yarns are needed for the weaving operation.

The twisting and plying operations vary the yarn strength, diameter, and flexibility and are important steps in producing the variety of fabrics which composite fabricators require.

2.4.3 Chopped strands

Continuous rovings or strands can be chopped into short lengths, usually between 3 to 50 mm long. Chopped strands are available with different sizings for compatibility with most plastics, the amount and type of size having a major influence on the integrity of the strand before and after chopping. A strand of high integrity is termed hard, and strand which separates more easily as soft. They can be used in a spray-up process, where chopped strands are sprayed simultaneously with liquid resin to build up reinforced plastic parts on a mould [26].

Chopped fibres are widely used as reinforcement in the injection moulding industry. The chopped glass and resin may be dry blended or extrusion compounded in a preliminary step before moulding, or the glass may be fed directly into the moulding machine with the plastic resin. Hundreds of different parts for many applications are made in this manner. Chopped glass may also be used as reinforcement in some thermosetting applications, such as sheet and bulk moulding compounds.

2.4.4 Mats

A mat is a blanket of chopped strand or of continuous strands laid down as a continuous thin flat sheet. The strands are evenly-distributed in a random pattern, and are held together by adhesive resinous binders or mechanically bound by needling. A chopped-strand mat is formed by randomly depositing chopped fibres onto a belt or chain and binding them with a chemical binder. Continuous-strand mat is formed in a similar manner, but without chopping, and less binder is usually required because of increased mechanical entanglement, which provides some inherent integrity [27].

The reinforcing ability of continuous-strand and chopped-strand mat is essentially the same, but they have different handling and moulding characteristics. Continuous-strand mat can be moulded to more complicated shapes without tearing. Needled mat, which has some fibres vertically oriented, is softer and more easily draped than non-needled mat, and therefore generally used only where reinforcement conformability is a particular requirement.

Reinforcing mats are distinguished by the binder used to hold them together, where binders may have high or low solubility. Solubility designates the rate of binder dissolution in the liquid resin matrix. Mats with a high-solubility binder are used in hand lay-up processes or wherever rapid wetting is important. Mats with a low solubility binder are used in press moulding or wherever the flow of the liquid matrix resin may wash away or disrupt the strands, leaving resin-rich areas.

Mats are used in hand lay-up, press moulding, bag moulding, autoclave moulding, and in various continuous impregnating processes.

2.4.5 Woven rovings

Woven rovings are produced by weaving fibre glass rovings into a fabric form. This yields a coarse product that is used in many hand lay-up and panel moulding processes to produce fibre-reinforced polymers. Many weave configurations are available, depending upon the requirements of the laminate. Plain or twill weaves provide strength in both directions, while a unidirectionally-stitched or knitted fabric provides strength primarily in one dimension. Many new woven roving fabrics are currently available, including biaxial and triaxial weaves for special applications (see **Figure 2.18**).

Woven rovings are heavier and thicker than woven yarn fabrics since rovings are heavier than yarns. Woven rovings typically weigh 40 to 140 g/m^3 and have thicknesses of 0.5 to 1.3 mm. They are usually provided in a plain weave, although special weaves have been developed. Woven rovings are usually moulded by hand lay-up. Typical applications include boats and cargo containers [26].

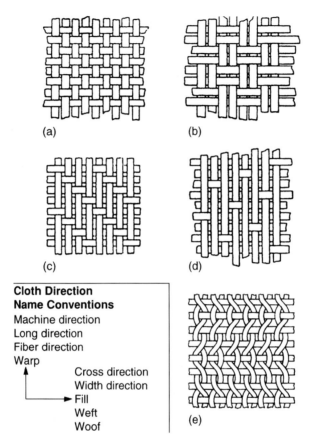

Figure 2.18 Common weave patterns (a) box or plain weave, (b) basket or twill weave, (c) crowfoot or five-harness satin weave, (d) long-shaft (satin) or eight-harness satin weave, (e) leno weave

2.4.6 Woven yarns

Woven yarns are manufactured by interlacing warp (lengthwise) yarns and fill (crosswise) yarns on conventional weaving looms. They are woven into a variety of styles, which permit quite exact control over thickness, weight, and strength. The principal factors that define a given woven yarn fabric style are fabric count, warp yarn, fill yarn, and weave [28].

The fabric count identifies the number of warp and fills yarns per inch. Warp yarns run parallel to the machine direction, and fill yarns are perpendicular. Fabric count, plus the properties of the warp and fill yarns used to weave fabrics, are the principal factors which determine fabric strength and drapeability.

The weave of a fabric refers to how warp yarns and fill yarns are interlaced. Weave determines the appearance and some of the handling and functional characteristics of a fabric. Among the popular weave patterns are plain, twill, crowfoot satin, long-shaft satin, leno, and unidirectional (see **Figure 2.18**). Plain weave is the oldest and most common textile weave. One warp end is repetitively woven over one fill yarn and under the next (**Figure 2.18a**). It is the firmest, most stable construction, providing porosity and minimum slippage. Strength is uniform in both directions. Twill weaves have one or more warp ends passing over and under two, three or more fill picks in a regular pattern (**Figure 2.18b**). Such weaves drape better than a plain weave. In the crowfoot and long-shaft satin weaves, one warp end is woven over several successive fill yarns, then under one fill yarn (**Figure 2.18c, d**). A configuration having one warp end passing over four and under one fill yarn is called a five-harness satin weave (**Figure 2.18c**). Similarly, in an eight-harness satin fabric, one warp end passes over seven fill yarns, then under one fill yarn (**Figure 2.18d**). The satin weave is more pliable than the plain weave. It conforms readily to compound curves and can be woven to a very high density. Satin weaves are less open than other weaves; strength is high in both directions. The leno weave has two or more parallel warp ends interlocked (**Figure 2.18e**). It tends to minimise sleaziness (flimsiness). Unidirectional weave involves weaving a great number of larger yarns in one direction with fewer and generally smaller yarns in the other direction. Such weaving can be adapted to any of the basic textile weaves to produce a fabric of maximum strength. Other weaves include basket, semi-basket, mock-leno, and high-modulus weaves.

Woven yarn glass fabrics are used to make a great variety of consumer products, e.g., boats, aircraft, pipe, tanks, electrical laminates, and ballistic armour.

2.4.7 Braids

Braiding is a mechanised textile process in which a mandrel is fed through the centre of the braiding machine at a uniform rate. Fibres from moving carriers on the machine braid about the mandrel at a controlled angle. The machine operates like a maypole, with carriers working in pairs to accomplish the over-under braiding sequence (**Figure 2.19a**) [27].

Since a braiding machine has many carriers, complete coverage of the mandrel can be achieved during one pass using an interwoven layer consisting of rovings lying at plus and minus the braiding. This is compared with multi-circuit helical filament winding, in which many circuits are required before a single layer is completed.

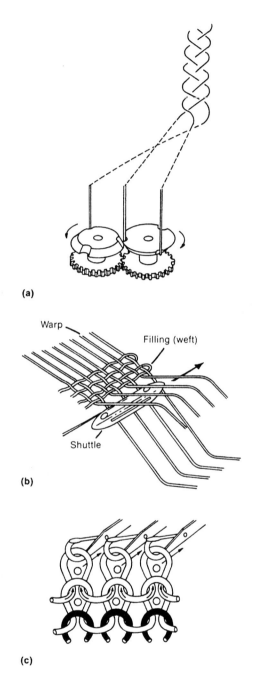

Figure 2.19 Fabric techniques (a) braided, (b) woven, (c) knitted

(Reproduced with permission from F.K. Ko in Engineered Materials Handbook, Volume 1, Composites, Ed. T.J. Reinhart, 1987, 520, Figure 1. ©1987, ASM International)

Depending upon the thickness of the roving used and the number of carriers, the resulting material is usually less closely interwoven than a standard fabric, comparable to so-called woven roving but more tightly interwoven than filament-wound material. It is difficult to braid an untwisted roving, although the difficulties are more severe with aramid and graphite fibres than with glass [28].

Since conventional braiding techniques were developed for textiles, there is no standard technique for applying the resin, as there is in filament winding. It may be applied after a braided layer is completed on the mandrel, although this step slows down the process. An alternative method of resin application involves passing the fibres over a porous ring, which delivers a required amount of resin to each roving. This method provides good fibre wetting and uniform resin content.

Because braiding is a rapid reinforcement-forming process, it produces a strong, interwoven tubular or flat structure from glass, carbon, or aramid yarns. The braids can be laid up over a mandrel wet and autoclave cured, or laid up dry and finished by resin transfer moulding (RTM).

2.4.8 Knits

Knitting is a process of interlooping chains of rovings or yarns (see **Figure 2.19c**) [27]. Knitting does not crimp the roving or yarn as weaving does, and higher mechanical properties are often observed in the reinforced product. Knitted fabrics are easy to handle and can be cut without falling apart.

The basic types of knits are 0°, or warp unidirectional, in which fibres run along the length of the roll; 90°, or weft unidirectional, in which fibres run along the width of the roll; bidirectional, with fibres at 0/90°, +45/-45°, or other angles; triaxial, with fibres in three directions, +45/-45/0° or +45/-45/90°; and quadraxial, with fibres at +45/-45/0/90° [28].

Knitted fibres are most commonly used to reinforce flat sections or sheets of composites, but complex 3-dimensional preforms have been created by using prepreg yarn. Broad goods (fibre woven to form fabric up to 127 cm wide) have been processed into ply sets (fabric consisting of one or more layers) and preforms as well. These materials have axially-oriented continuous fibres in the desired direction. Multiple plies of these fibres in the x-y direction are stitched in the z direction to achieve the balance of drapeability and reinforcement properties desired in processing and in performance of the finished parts.

2.4.9 Preforms

Preforms are the cut and shaped fibre forms for reinforcement. In preforming processes, two-dimensional materials, e.g., mats, woven yarns, prepregs etc., are converted into three-dimensional shapes ready to be used in moulding processes.

Materials may be laid up in any orientation, using as many plies as necessary to achieve the desired thickness. Ply drop-offs can be handled with ease, and foam cores or metal inserts can be incorporated. Once the patterns are cut, they are assembled into the shape of the part being produced, including the positioning of inserts or cores. They are then held together by one of three methods: sewing (by hand or machine), stitching or quilting, or the use of an uncatalysed solid epoxy resin that is either sprayed between the plies as a molten material, or dusted on and the heat set later. Usually only 2%–4% epoxy is needed to hold the preform shape.

Several techniques can be used to speed up the production process for preforms. One is the use of pre-stitched blankets containing the needed plies in the proper orientations.

2.5 Fibre-matrix combination forms

Combination of the reinforcing fibres into the polymer matrix either occurs during the shaping process or before the process. In the first case, the starting materials arrive at the fabrication operation as separate constituents and are combined into the composite during shaping. Examples of this case are filament winding and pultrusion. The starting fibre reinforcement in these processes consists of continuous fibres. In the second case, the two constituent materials are combined into some preliminary form that is convenient for use in the shaping process. Nearly all the thermoplastics and thermosets used in the plastic-shaping processes are polymers combined with fillers. The fillers are either short fibres or particulates (including flakes).

Starting form can be considered to be prefabricated composites that become primed for use at the shaping process. There are two types of shaping forms: moulding compounds and prepregs.

2.5.1 Moulding compounds

Moulding compounds are similar to those used in plastic moulding. They are designed for use in moulding operations and so they must be capable of flowing, at least to some degree. Most moulding compounds for composite processing are thermosetting polymers. They have not been cured prior to shaping processes. Curing is done during and/or after final shaping. Moulding compounds consist of the resin matrix with short, randomly-dispersed fibres. There are three widely used forms [25].

2.5.1.1 Sheet moulding compounds (SMC)

SMC are a combination of thermosetting polymer resin, fillers and other additives, and randomly-oriented chopped glass fibres, all rolled into a sheet of typical thickness of 6.5 mm. The most common resin is unsaturated polyester; fillers are usually mineral powders such as talc, silica, limestone; and the glass fibres are typically 12 to 75 mm in length and account for about 30% of the SMC by volume. SMC are very convenient for handling and cutting to proper size as moulding charges. They are generally produced between thin layers of polyethylene to limit the evaporation of volatile compounds from the thermosetting resin (**Figure 2.20**) [27]. The protective coating also improves surface finish on subsequent moulded parts.

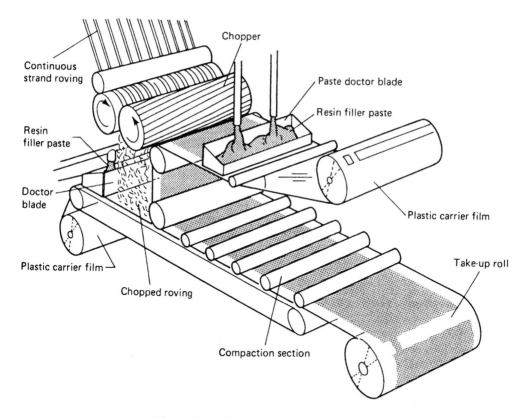

Figure 2.20 SMC processing machine
(Reproduced with permission from J.J. McCluskey and F.W. Doherty in Engineered Materials Handbook, Volume 1, Composites, Ed. T.J. Reinhart, 1987, 157, Figure 1.
© *1987, ASM International)*

2.5.1.2 Bulk moulding compounds (BMC)

BMC consist of similar ingredients as in SMC, but the compounded polymer is in billet form, rather than sheet. The fibres in BMC are shorter, typically 2 to 12 mm, because greater fluidity is required in the moulding operations for which these materials are designed. The billet diameter is usually 25 to 50 mm. The process for producing BMC is similar to that for SMC, except that extrusion is used to obtain the final billet form. BMC is also known as dough moulding compound (DMC).

2.5.1.3 Thick moulding compounds (TMC)

TMC are the new fibre-reinforced polymers, suited for compression, injection, and transfer moulding, and can be processed on the same equipment as SMC and BMC materials. TMC composites can be produced up to 50 mm, in contrast to the 6.5 mm maximum thickness of SMC. Glass fibre length in TMC is 6 to 50 mm. During fabrication, complete wet-out of resins, fillers, and reinforcement fibres is achieved, resulting in improved mechanical properties and reduced porosity. Low porosity means that TMC affords better surface qualities on moulded products than either SMC or BMC, which also means that less re-work is needed on TMC end products.

2.5.2 Prepregs

Prepregs are material forms consisting of continuous unidirectional or woven fibres pre-coated with a controlled quantity of an uncured catalysed resin matrix material [29]. They are supplied in roll or sheet form, ready for immediate use at a composite manufacturing facility, and are widely used in the aerospace and other industries for high-performance structural applications. Composite parts are manufactured from prepreg by the basic steps of lay-up, cure, and finishing.

At the present time, the principal fibres used are aramid, carbon, and glass; with epoxy, bismaleimide, phenolic, or polyimide resin as the matrix. However, almost any fibre/matrix combination can be produced.

The predominant methods of prepreg production are via a hot melt or a solvent impregnation. In the hot melt method, a film, (or films), of formulated resin at a controlled weight is impregnated onto the fibre form using heat and pressure (**Figure 2.21**) [30]. For solvent impregnation, the fibre form is passed through a solution of formulated resin, calendered to produce the desired resin content, and then sent through a heated oven to remove the solvent [30].

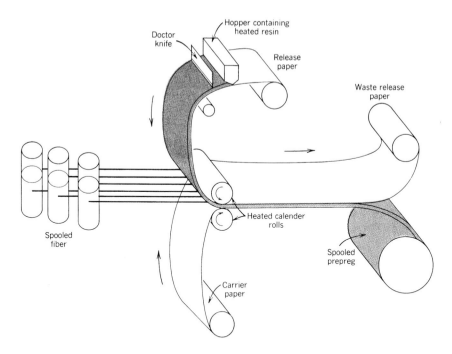

Figure 2.21 Prepreg tape production
*(Reproduced with permission from W.D. Callister, Materials Science and Engineering:
An Introduction, 5th Edition, 1999, 548, Figure 17.13. © 1999, John Wiley
& Sons, Incorporated)*

The resin flow, tack, drapeability, and gel time of the prepreg can be tailored to meet end-user requirements. Resin flow is a measure of resin movement during the cure process. Flow requirements are determined by the type of process used. Tack is a measure of the ability of a ply of prepreg to stick to another ply and to the tool surface. This is important in the lay-up processes. Drapeability is a measure of the ability of the prepreg to conform to contoured tool surfaces without fibre damage, and is also important in the lay-up processes. Gel time is a measure of the time taken for the resin to reach an immobile state at a given temperature. This is a key parameter associated with the cure process used.

The properties of the major commercial prepreg types are as follows [27]:

i)   Unidirectional prepreg tapes

They are available in widths ranging from 25 to 1500 mm, with finished cured ply thickness ranging from 0.05 to 0.5 mm. As the fibres all run in one direction, and are not crimped as in the woven fabric prepreg, the tape gives the best translation of the reinforcing fibre properties. However, the resultant cure part cost is higher, as there is an increased labour cost associated with the lay-up procedure compared to fabric. Because of this, there is an increased use of robotic methods for lay-up of tape to try to reduce labour.

ii) Woven prepregs

They are usually supplied as broadgoods with widths up to 1800 mm and cured ply thickness of 0.01 to 0.8 mm. The weave styles can be varied to fit particular end needs. Typical examples are balanced weave, 45°, or bias weave and 90° unidirectional. Mechanical properties or fibre strength translation depend on the weave style, e.g., a square weave gives lower translation than an 8-harness satin because of the distance between fibre interweaves. Part lay-up costs are lower as typically fewer plies are used. However, there is a trade-off as the cost of woven prepreg material is higher than tape.

iii) Prepreg Tows

They are supplied as individual pre-impregnated fibre bundles on spools in the same type of package as the original dry fibres. They are used for filament winding, where they offer the potential for low-cost manufacturing, using high-performance matrix resins compared to typical wet wind systems. They are also used as local reinforcements or to fill in layed-up parts such as beams. Properties of the prepreg tows are similar to unidirectional tape material.

Prepregs are used extensively in the aerospace industry; in military and commercial aircraft, satellites, and missiles where weight and performance are critical factors. For example, most aircraft interiors are produced using prepregs of glass and aramid, and there is increased use for structural components where metal is replaced by carbon composites. Other commercial applications include the recreational industry, where prepregs are used in tennis racquets and golf clubs.

References

1. L.A. Pilato and M.J. Michno, *Advanced Composite Materials*, Springer-Verlag, New York, USA, 1994.

2. R. Fasth and C.H. Eckert, *ChemTech*, 1988, **18**, 7, 408.

3. J.K. Gilham, *Makromolekulare Chemie Makromolecular Symposia*, 1987, **7**, 67.

4. J.K. Gilham in *Advances in Polymer Science*, Ed., K. Dusek, Springer-Verlag, Heidelberg, 1986, Volume 78, 83.

5. J. McGrath, *SAMPE*, 1987, **32**, 1276.

6. *Concise Encyclopedia of Polymer Science and Engineering*, Ed., J.I. Kroschwitz, Wiley Interscience, New York, 1998, 818.

7. T.J. Reinhart in *Engineering Materials Handbook: Composites*, ASM International, Ohio, USA, 1987, Introduction.

8. T.L. Vigo and B. Kinzig, *Composite Applications*, VCH Publishers Inc., New York, USA, 1992.

9. J.C.D.A. Lewis and M. Singh, *TMS Materials Week*, 1993, October.

10. R.S. Bauer in Chemistry and Properties of High Performance Composites Workshop sponsored by ACS Polymer Division, 1988.

11. G.R. Almen, R.K. Maskell, V. Malhorta, M.S. Seftan, P.T. McGroil and S.P. Wilkinson, *Society Advanced Materials Processes Engineering Series*, 1990, **35**, 1, 419.

12. P.M. Hergenrother and N.J. Johnston, *Polymer Material Science Engineering Procedures*, 1988, **59**, 697.

13. B.A. Zhubanov and B.K. Donenov, *Polymer Science*, 1992, **34**, 10, 819.

14. R.P. Sheldon, *Composite Polymeric Materials*, Applied Science Publishers, London, UK, 1982.

15. J.C. Watson in *Composites Handbook*, Ed., T. J. Reinhart, ASM International, Ohio, USA, 1987, Volume 1.

16. W.D. Bascom in *Composites Handbook*, Volume 1, Ed., T.J. Reinhart, ASM International, Ohio, USA, 1987.

17. E. Fitzer in *Carbon Fibers and Composites*, Eds., J.L. Figueirado, C.A. Bennado, R.T.K. Baker and K.J. Huttinger, Nato ASI Series E, Applied Sciences, 1989, Volume 177, 3.

18. T. Schoenberg, in *Handbook of Composites*, Volume 1, Ed., T.J. Reinhart, ASM International, Ohio, USA, 1987.

19. K.K. Chawla, *Composite Materials: Science and Engineering*, 1st Edition, Springer-Verlag, New York, USA, 1987.

20. C.T. Herakovich, *Mechanics of Fibrous Composites*, J. Wiley and Sons, New York, USA, 1998.

21. K. Suzuki, Y. Mukoyama and T. Ito, inventors; Hitachi Chemcial Company, assignee; US Patent 5874518, 1999.

22. T. Takatani and H. Ishida, inventors; General Electric Company, assignee; US Patent 5872187, 1999.

23. J-P. Cougoulic, inventor; no assignees; US Patent 5872159, 1999.

24. M. Hofsass, inventor; M. Hofsass, assignee; US Patent 5877671, 1999.

25. M.P. Groover, *Fundamentals of Modern Manufacturing*, Prentice Hall, New Jersey, USA, 1996.

26. M.M. Schwartz, *Composite Materials*, Volume 2, Prentice Hall, New Jersey, USA, 1997.

27. *Engineered Materials Handbook, Volume 1, Composites*, ASM International, Ohio, USA, 1987.

28. M.M. Schwartz, *Composite Materials Handbook*, McGraw-Hill, New York, USA, 1984.

29. *Plastics Handbook*, Modern Plastics Magazine, McGraw-Hill, New York, USA, 1994.

30. W.D. Callister, *Materials Science and Engineering: An Introduction*, 5th Edition, John Wiley & Sons, Inc., New York, USA, 1999.

3 Open Mould Processes

C. Kaynak and T. Akgül

3.1 Introduction

Only one mould surface is used in open mould process. This single mould represents either the positive (male plug) or negative (female cavity) surface as shown in **Figure 3.1** [1]. In order to produce large components and structures, (for instance swimming pools, boat hulls, etc.) very large moulds are usually used. The main matrix materials used are thermosetting resins of epoxy and polyester, while E-glass fibres are the most widely used reinforcement material. Depending on the desired thickness, matrix resins and reinforcement fibres are applied to the mould surface layer by layer. The fibre reinforcement can be used in the form of mats, woven roving or yarns. The use of prepregs may simplify the laying process. After the lay-up process, curing treatment will be necessary for rigid thermoset matrices. Depending on the type of resin used, little or no pressure will be necessary during curing.

Open mould processes have several advantages over closed mould composite manufacturing processes. Since a single mould is used in open mould processes, mould costs will be much less than using two moulds in the closed mould processes. Another

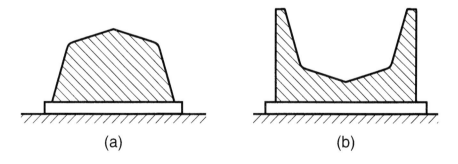

(a)　　　　　　　　　　(b)

Figure 3.1 Types of open mould: (a) positive and (b) negative
(*Reproduced with permission from M.P. Groover, Fundamentals of Modern Manufacturing, Materials, Processes and Systems, 1996, 378, Figure 17.3.*
© *1996, John Wiley & Sons Incorporated*)

advantage is that very large and complex structures may be produced in open mould processes, which is difficult in closed mould processes. Depending on the component to be produced, a variety of materials (e.g., metals, plaster, woods, fibre reinforced composites) are available for cheaper open moulds, whereas expensive metallic moulds should be designed for closed mould processes. Therefore, it may be concluded that open mould processes have better design flexibility compared with closed mould processes.

However, open mould processes have some disadvantages as well. Firstly, only one surface of the product will be finished and smooth. This is because the other surface will be not in contact with the open mould surface. Moreover, to achieve a good surface finish on at least one surface of the component, the surface of the open mould must also be very smooth. The second disadvantage of open mould processes is that they are very labour intensive. Therefore, for the production of components with higher quality, the personnel working in the process should be adequately skilled.

There have been many advances in the automation of open mould processes, which helps to solve the skilled personnel problem. Automation in open mould processes is increasing not only the quality of the product, but also the number of the parts manufactured per unit time. Another disadvantage of open mould processes is the much longer curing periods required compared with other methods. Normally, application of heat will decrease curing time. However, it is difficult to heat treat components which are very large.

Open mould processes are usually classified according to the methods of resin and reinforcement application to the mould, or according to the curing methods. If the matrix and reinforcement is applied by hand, then it is named hand lay-up, if it is by a spray gun, then it is called spray-up. Similarly, if the curing is accomplished in a bag, then it is called bag moulding, if it is performed in an autoclave, then it is termed autoclave moulding, etc. However, in order to use the advantages of each method, generally two or more of these methods are combined during manufacturing. These methods will be explained briefly in this chapter, with more emphasis being placed on autoclave moulding, due to its technological importance.

3.2 Wet lay-up processes

3.2.1 Hand lay-up

Hand lay-up, which is one of the oldest open mould composite processing methods, was first used to manufacture boat hulls in the middle of the last century. Hand lay-up is a composite laminating process in which layers of resin and reinforcement are applied manually onto an open mould surface one by one until the desired thickness of the component is obtained.

There are five main steps in lay-up procedures: cleaning, gel coating, laying-up, curing, and part removal. These steps (see **Figure 3.2** [1]), are generally also required for other open mould processes. The main differences between each open mould process takes place during the third (laying-up) and fourth (curing) steps. After these steps, the outside edges of the finished components should be trimmed and machined.

The first step of the hand lay-up process is cleaning the surface of the mould, followed by the application of a release agent for easy part removal. In the second step, a thin gel coating will be applied to the outside surface of the moulding, if the surface quality of the product is important. The gel coating resin is usually pigmented (in order to colour the surface), and applied to the mould by using a spray gun.

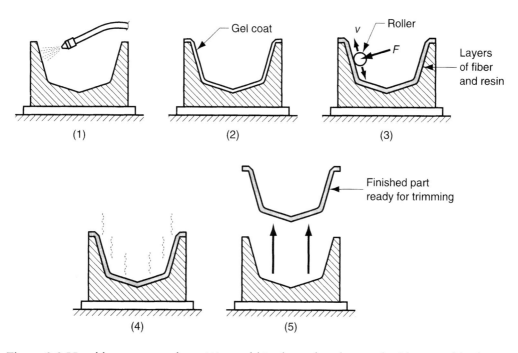

Figure 3.2 Hand lay-up procedure: (1) mould is cleaned and treated with a mould release agent; (2) a thin gel coat (resin, possibly pigmented to colour) is applied, which will become the outside surface of the moulding; (3) when the gel coat has partially set, successive layers of resin and fibre are applied, the fibre being in the form of mat or cloth; each layer is rolled to fully impregnate the fibre with resin and remove air bubbles; (4) the part is cured; and (5) the fully hardened part is removed from the mould.

The third step begins when the gel coat is partially set. In this step of the hand lay-up, as the name suggests, resin and fibre forms are manually applied to the open mould as successive layers. Each layer is consolidated by a roller, ensuring that the resin impregnates the fibre and that any air bubbles that are present are removed. The fourth step is the curing stage, which is applied in order to harden the part. This will be discussed in section 3.3.3. In the fifth and final step, the component is removed from the mould, and is now ready for trimming and other surface finishing processes.

Various forms of fibre reinforcement may be used in hand lay-up: chopped strand mats, woven roving, woven yarns, or cloths. Usually these forms are supplied in rolls, which are unrolled during lay-up. Unrolled fibre forms may be cut to size by using templates or patterns. After cutting, reinforcement layers are positioned on the prepared mould, followed by wetting with the resin mixture.

Resins are generally catalysed by hardeners and accelerators. For the application of catalysed resins to wet out the fibre layers, two methods are used: one involves a brush while the other employs a spray gun with a pumping system. Today, the brush method is generally used in small manufacturing facilities and for low production rates. However, many manufacturers prefer to use spray guns and pumping systems in order to achieve rapid and consistent resin mix delivery for better layer wet out.

In order to compact successive layers in the laminates, rollers, squeegees, or stipple brushes are used. Rolling will also remove air bubbles trapped in the laminate that would otherwise decrease the mechanical strength of the final product. Depending on the thickness of the product, alternating layers of resin and reinforcement may be added. After each new layer has been added, it is rolled out manually.

This approach, i.e., the application of alternating layers of resin and fibre forms, is also called wet lay-up. An alternative approach to wet lay-up is the use of prepregs, which are produced before the process, outside the mould, and easily laid onto the open mould surface. There are several advantages of using prepregs in lay-up, including closer control of the fibre-resin mixture, and more efficient addition of new layers onto the laminate.

There are several materials used in the production of moulds for hand lay-up processes, such as metals, glass fibre reinforced plastics, or plaster of paris. The factors that affect the choice of the mould material include the number of the products to be produced, the surface quality of the product required, as well as several other technical and economic factors.

If only one part is to be produced, (for example in prototyping), then the mould can be made of plaster of paris. It is better to use glass fibre reinforced plastic moulds for the production of medium size quantities. Of course, for very high production rates and sizes the best choice would be metallic moulds, due to their higher durability. Another advantage of metallic moulds is their high thermal conductivity. Higher conductivity will improve the heating process during curing or it will aid heat dissipation from the composite laminate during room temperature curing.

Hand lay-up is especially suited for very large components, but for lower production quantities. There are several advantages in low volume production, such as flexibility in mould design, use of cheaper mould materials, almost unlimited mould size, lower cost of other equipment, etc. It is also easy to repair damaged parts or rejects during the early periods of the production process. Due to the possibility of economical design changes, hand lay-up is also suitable for complex works.

Today, hand lay-up is extensively used in the production of boat hulls, large container tanks, stage props, swimming pools, as well as for various structural plates and panels. Hand lay-up is also used for some automotive parts. However, it is not economical due to its lower production rate.

3.2.2 Automation in lay-up

The best way to automate the hand lay-up process is to accelerate the laminating operation in the third step. Therefore, equipment manufacturers have developed automated tape-laying machines for this purpose. The procedure is almost the same with hand lay-up. The process begins with the coating of the outer surface of the mould with a release agent and, if required, a gel coating may also be applied. The difference is the use of an automated tape-lying machine in the third step, which is followed by the same steps of curing and part removal.

Automated tape-laying machines may use rolls of chopped strands, woven roving and yarns, or prepreg tapes. Initially, rolls of these fibre forms are mounted on the machine, and then they are dispensed through a catalysed matrix resin for fibre wetting. Then, these wetted fibre forms pass through rollers. The gap between the rollers is adjusted according to the laminate thickness required. This gap is important for the determination of the resin amount desired, and when the material passes through the rollers, excess resin is also removed. Then, the material is positioned onto the mould just after leaving the machine.

Each lamination is placed by following a series of back-and-forth passes across the mould surface until the layer is completed. There are some factors affecting the rate of automated tape-laying machines. These include the desired laminate thickness, the looseness/tightness of the woven forms, and also the resin type used for fibre impregnation.

The automated tape-laying machines are efficient in controlling the matrix resin/reinforcement fibre content according to the desired fibre/resin ratios. Hence, the product quality and consistency is much better compared to manual hand lay-up.

Automated tape-laying machines are available in various sizes, from a small portable model to a very large model used for boat hulls. They are capable of multi-directional motion, full rotation, and vertical telescoping. Vertical telescoping is where vertical tubular parts or colliding parts of the machine press or drive together so that one slides into another like the sections of a folding telescope. These machines have been developed mainly by the aircraft industry to decrease labour costs and also to achieve the highest quality and uniformity in the manufactured components. The only disadvantage of these computer numerically-controlled machines is that they must be programmed, which is time consuming.

Automated tape-laying machines are especially suitable for large and simple geometries. The main application area is the manufacture of aircraft body components.

3.2.3 Spray-up

Spray-up is another alternative of the third step of the open mould wet lay-up process. It helps to automate the application of the matrix resin and reinforcement fibre layers, thereby reducing the time period required in the manual lay-up procedure.

In this method, as illustrated in **Figure 3.3** [1], chopped fibres, together with the liquid matrix resin, are sprayed onto an open mould surface until the desired thickness of the composite lamination is obtained by successive layers. Spray-up may also be used in applying the gel coat to the mould surface before hand lay-up and spray-up processes.

The equipment necessary in the spray-up process includes: a spray gun, a glass fibre chopper attachment and a pumping system. The chopper attachment delivers continuous roving and cuts them into short fibres. Generally, chopped fibres have a length of between 25 and 75 mm, and they are added to the matrix resin stream as it exits the spray gun nozzle. This type of mixing leads to random orientation of the fibres in the layer, whereas in hand lay-up fibres may be oriented.

The composite manufacturing industry uses four main types of spray gun. They are internal mix with air, airless internal mix, external mix with air and airless external mix [2].

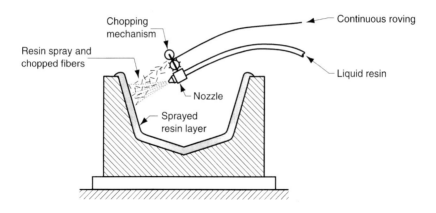

Figure 3.3 Spray-up method

(Reproduced with permission from M.P. Groover, Fundamentals of Modern Manufacturing, Materials, Processes and Systems, 1996, 380, Figure 17.5. © 1996, John Wiley & Sons Incorporated)

- Airless Internal Mix: In this type of spray gun, the matrix resin and the catalyst first flow in as two separate streams into a single chamber in the gun head. Then, the resin and the catalyst are combined and mixed in this chamber, after which they are sprayed together through the gun nozzle. The advantage of the internal mix spray gun is that it provides a proper mixing of resin and catalyst, which results in higher quality products. Proper mixing also decreases the overspray during the process, which reduces the amount of wasted resin. The disadvantage of internal mix method is that spray gun equipment requires a solvent flushing system to clean the equipment when the production is interrupted.

- Airless External Mix: In this type of spray gun, resin and catalyst are combined and mixed in mid-air, not in a closed chamber. In external mixing, one spray gun head or two separate gun heads may be used. Therefore, the advantage of external mix guns over internal mix guns is that they do not require a solvent flushing system, which leads to less complex operation. Solvent flushing will only be needed to clean the equipment after longer manufacturing periods. However, the disadvantage of the external mix method is that it can not mix the resin and catalyst as well as the internal mix method, leading to less efficient use of catalyst and overspray.

- Internal and External Mix with Air: If air is injected into the resin stream or catalyst stream in the spray gun, then the spraying operation becomes simpler. However, there are several disadvantages of internal or external mix with air. These include: a great deal of overspray, extensive fuming, an increased amount of air trapped in the laminate, leading to higher porosity, etc.

After the application of the catalysed resin mixture and chopped fibres through the spray gun onto the open mould surface, rolling will be necessary to compact the laminate as in hand lay-up. Manufacturers use hand rollers or automated rolling systems. Usually hand rolling is adequate for smaller parts, whereas automated rolling systems are preferred for larger parts with flat surfaces.

Spray-up is used in the manufacture of various components, including: boat hulls, shower stalls, bathtubs, automotive body parts, tanks, containers, structural panels, furniture, etc. These components may also be produced by the hand lay-up method.

The difference between spray-up and hand lay-up is the strength of the product manufactured by these methods. In spray-up, short fibres are randomly oriented, while in hand lay-up fibres may be continuous and aligned, which will result in higher strength. Another difference between these methods is the amount of fibre percent that may be used. Due to the shortcomings of the spraying and mixing process in the spray-up method, fibre content is limited to 30%–35%, whereas in hand lay-up it may be up to 65%–70%.

However, there are several advantages of the spray-up method over the hand lay-up method. First, spray-up is much quicker and less labour-intensive than hand lay-up. Secondly, simple and cheaper open moulds may also be used. Moreover, on-site manufacturing of the components will be possible by using portable spray-up equipment, which is available. Therefore, spray-up may be considered as the most versatile open mould processing method.

3.2.4 Automation in spray-up

In the spray-up process, spraying onto an open mould surface is generally performed manually by an operator who uses a hand-held portable spray gun. This kind of manual spraying must be applied by a competent worker in order to operate and maintain the equipment, and also to produce the proper lamination for the component being manufactured. Therefore, due to the dependence of the quality and consistency of the laminate on operator skill, components produced in this way will vary in weight and quality.

Automation in spray-up will prevent these negative influences on the manufactured components. Automation or mechanisation in spray-up is achieved by using a computerised machine in which the path of the spray gun is programmed and controlled. In this method, the spray gun is either attached to an operator-controlled machine or it is operated by a robot. Therefore, automation in the spray-up process reduces labour force and results in much more consistency and quality in the product compared to manual spraying.

Automation also makes spray-up an environmentally-friendly process. However, if the spray-up equipment is inadequate, or is operated by non-competent personnel, then more fuming will be created in the work place. Fuming and other volatile emissions from the liquid resins are hazardous to health. Therefore, using automated equipment, in the absence of a worker, will be advantageous.

3.3 Bag moulding and curing processes

The purpose of the bag moulding process in wet lay-up techniques is to compact the lamination on the open mould surface. Another function is to drive out volatiles by applying some kind of pressure to the uncured matrix resin. Therefore, in hand lay-up and spray-up processes, bag moulding can be used as a supplementary process to curing. As illustrated in **Figure 3.4** [1], there are two basic methods used in industry: vacuum bag moulding and pressure bag moulding.

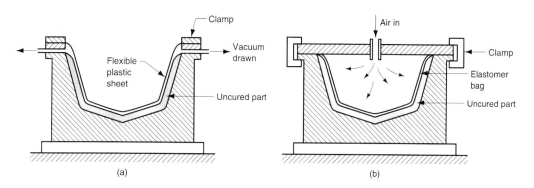

Figure 3.4 Bag moulding procedures: (a) vacuum bag moulding and (b) pressure bag moulding

(*Reproduced with permission from M.P. Groover, Fundamentals of Modern Manufacturing, Materials, Processes and Systems, 1996, 382, Figure 17.7. © 1996, John Wiley & Sons Incorporated*)

3.3.1 Vacuum bag moulding

In vacuum bag moulding, the laminated component produced by hand lay-up or spray-up on the open mould is covered by a polymeric sheet. The sheet must be flexible and non-adhering to the laminate: polyvinyl alcohol or Nylon is usually chosen. After sealing the edges of the covered mould, a vacuum is drawn to apply pressure by the plastic bag against the laminated component during curing, as shown in **Figure 3.4**.

Compared to lay-up processes without bag moulding, the vacuum method provides higher reinforcement content and better adhesion between layers. In order to accelerate curing, heating may be applied in the vacuum bag technique. Compared with pressure bag moulding, the limitation of the vacuum bag method is the maximum amount of the pressure supplied, which can not exceed 98 kPa [3].

3.3.2 Pressure bag moulding

The difference between pressure bag and vacuum bag moulding is the type of pressure applied. In the vacuum bag method, pressure is supplied due to the vacuum drawn between the plastic bag and the laminate. However, in the pressure bag technique, a positive air pressure is applied. In this method, positive pressure is supplied by blowing air to inflate an elastomeric bag covering the laminate on the open mould surface (**Figure 3.4b**). Another difference is the magnitude of the pressure that may be applied, which can be $\sim$ 300 kPa in pressure bag moulding, while it is less than 98 kPa in the vacuum bag method.

Pressure bag moulding is also applied while the curing process proceeds. In order to accelerate curing, either external heating is added, or instead of air pressure, pressurised steam is blown onto the elastomeric bag [4]. Pressure bag moulding is especially suitable for complex hollow components, which normally need cores and inserts.

3.3.3 Curing

As stated in Section 3.2, curing is the fourth step in all wet lay-up processes, and is required for all thermosetting matrix resins to harden. During curing, the liquid or viscous form of the matrix resin transforms into a hardened rigid state due to the extensive crosslinking of the thermosetting polymeric structure.

Time, temperature and pressure are the three main parameters influencing the degree of crosslinking in the curing process. In thermosetting resins, curing will start at room temperature, which is rather slow. Increased temperature will decrease the time necessary for the curing process to be completed. However, in wet lay-up processes (including hand lay-up and spray-up), the moulds are usually very large in size. Hence, heating can be very difficult, and sometimes several days may be required for sufficient hardening before the removal of the products from the mould.

There are several heating techniques used in curing processes, such as oven curing, infrared curing, and autoclave curing [5]. In oven curing, heat is applied at controlled temperatures, and it is also possible to draw a partial vacuum in some curing ovens. A typical vacuum curing oven (a vessel that provides heat by convection), is a large metal, thermally-insulated, air-circulating oven with large doors at one or both ends. An example of standard size is 3 m high, 4 m wide, and 9 m long. Infrared heating is especially used when the laminate on the open mould is very large in size, and therefore impractical to heat in an oven.

In autoclave curing, heat is applied in an enclosed chamber, both under temperature and pressure control, which are the most important parameters, especially in bag moulding processes. An autoclave is a large cylindrical metal pressure vessel pressurised with air and/or CO_2. It is thermally insulated, with circulating hot air and a large circular door at one or both ends. A typical size is 4 m in diameter and 17 m long. Autoclave curing (which is discussed in detail in section 3.4) is widely used in the aircraft industry.

3.4 Autoclave moulding process

3.4.1 Autoclave cure systems

The autoclave moulding process is used mainly in the aerospace industry where high production rate is not an important consideration. It is used mainly for manufacturing composite parts, besides various other applications like the vulcanisation of rubber products.

An autoclave system is a pressure vessel where a complex chemical reaction occurs inside according to a specified schedule (cure cycle) in order to process a variety of materials. With the developments in materials and processes, autoclave operating conditions go up to 700 °C and 15 MPa. The materials processed in autoclaves include metal bonding adhesives, thermoplastic laminates, metal, ceramic and carbon matrix materials, as well as other aerospace and electronic components. In **Figure 3.5**, a typical autoclave system is depicted. The notable elements of an autoclave system are:

- a pressure-containing vessel
- a heat source
- a fan to circulate gas inside uniformly within the vessel
- a system to pressurise the vessel
- a system to apply a vacuum to parts covered by a vacuum bag
- a system to control operating parameters, and
- a system to load the moulds into the autoclave.

Figure 3.5 A typical autoclave system
(*Reproduced with permission of Scholz GmbH and Co KG, Coesfeld, Germany*)

The autoclaves are usually pressurised with nitrogen or carbon dioxide from a liquid storage tank and which is vaporised before use. Previously, autoclaves were pressurised with plant air, but this arrangement carried a fire risk. However, nowadays some companies prefer to use both air and nitrogen in a certain ratio for economical reasons.

While designing and fabricating pressure vessels, it is compulsory to comply with the American Society of Mechanical Engineers (ASME) requirements, and all vessels are tested according to those criteria.

Autoclaves are heated through a gas-fired heat exchanger system or an electrical heating system.

The control systems are generally computer-controlled, coupled with a manual back-up system. With the use of computer-control systems in production, better autoclave cure cycles are generated and monitoring the curing internally (and therefore obtaining an optimum cycle) becomes possible.

3.4.2 Materials to be used

The benefits of composite materials are now well known and a great variety of applications, ranging from industrial and sports and leisure to high performance aerospace components, indicate that composite materials have a promising future.

A composite is a combined material created by a synthetic assembly of two or more distinct materials—a selected filler or reinforcing agent and a compatible matrix binder (i.e., a resin)—in order to obtain specific characteristics and properties. The components of a composite do not dissolve or otherwise merge completely into each other, but however do act in concert. Properties of such a composite cannot be achieved by any of the components acting alone. [6]

Reinforcement types and matrix systems are described primarily in Chapter 2 of this handbook. In this section, particular attention will be given to defining the materials used in autoclave moulding.

When discussing the composite materials employed in autoclave moulding, we must mention prepreg materials. A load-carrying fibre structure is impregnated with resin and is ready for use in the component manufacturing process. The resin is partially cured to B-stage and supplied to the fabricator, who lays-up the finished shape and completes the cure with the application of heat and pressure. Prepregs are available in Unidirectional form or Fabric form as seen in **Figure 3.6**.

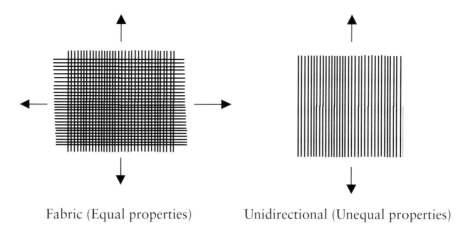

Fabric (Equal properties)   Unidirectional (Unequal properties)

Figure 3.6 Prepreg forms

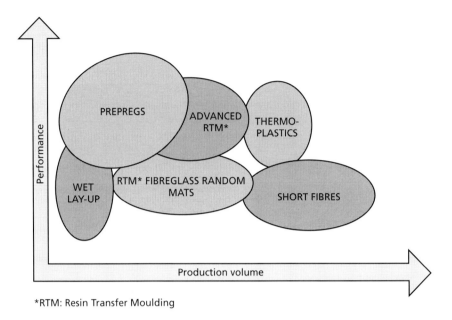

*RTM: Resin Transfer Moulding

Figure 3.7 Comparison of fabrication processes
(*Reproduced by courtesy of Hexcel Composites Ltd., Cambridge, UK*)

The state of prepreg technology in terms of performance and production volumes is compared with other fabrication processes in **Figure 3.7**.

3.4.2.1 Fibres

In a continuous fibre reinforced composite, fibres present virtually the entire load-carrying characteristics of the composite, the most important of which are strength and stiffness. Multiple fibres in a composite make it a redundant material because the failure of even several fibres results in the redistribution of load onto other fibres, rather than a catastrophic failure of the part. Fibre materials can be classified as:

- Glass

- Carbon

- Aramid

- Other organic fibres (polybenzimidazole, polyethylene, etc.)

- Boron

- Continuous silicone carbide (SiC)

- Aluminium oxide

3.4.2.1.1 Fibre and fabric properties

Reinforcement materials provide composites with mechanical performance, including excellent stiffness and strength, as well as good thermal, electric, and chemical properties. They also offer significant weight savings over metals. The range of fibres is extensive, and some fibres are compared in terms of some important material parameters in **Table 3.1**.

Table 3.1 Comparison of fibres				
Best ⟵ ⟶ Worst				
Cost	E-Glass	S-Glass	Kevlar	Graphite	Ceramic
Density	Kevlar	Graphite	S-Glass	E-Glass	Ceramic
Stiffness	Graphite	Kevlar	S-Glass	Ceramic	E-Glass
Heat	Ceramic	S-Glass	E-Glass	Kevlar	Graphite
Toughness	Kevlar	S-Glass	E-Glass	Ceramic	Graphite
Impact	Kevlar	S-Glass	E-Glass	Ceramic	Graphite

3.4.2.1.2 Different styles of fabrics

Fabrics consist of at least two threads, which are woven together: the warp and the weft. The weave style can be varied according to crimp and drapeability. Low crimp gives better mechanical performance because straighter fibres carry greater loads: a drapeable fabric is easier to lay-up over complex forms. There are four main weave styles (see **Figure 3.8**).

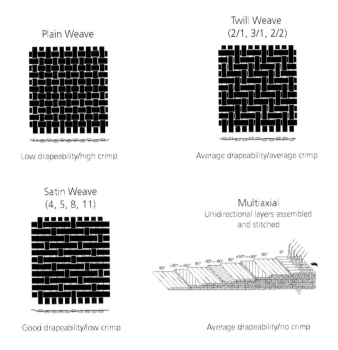

Figure 3.8 Weave styles

(Reproduced by courtesy of Hexcel Composites Ltd., Cambridge, UK)

3.4.2.1.3 Main factors affecting the choice of reinforcement

Reinforcements come in various forms, and each type offers particular advantages, as shown in **Table 3.2**.

	Reinforcement	Advantages	Applications
Unidirectional	Tape	High strength and stiffness in one direction Low fibre weights $\cong 100 \text{g/m}^2$	Sports goods Aircraft primary structures*
	Single Tow	Suitable for filament winding Very narrow width for accurate fibre placement (1 mm)	Pressure vessels Drive shafts Tubes
	Strips	High strength and stiffness in one direction High fibre weights $\cong 1,000 \text{ g/m}^2$ Economic processing	Yacht masts Leaf springs Skis
	Fabrics > 80% warp	For components requiring predominant strength and stiffness in one direction Good handling characteristics Weights from 160 to 1,000 g/m^2	Aerospace Industrial Sport and leisure
Woven Fabrics	Balanced Fabrics	Strength and stiffness in two directions Very good handling characteristics Good drape Choice of weave styles Possible to mix fibres Weights from 20 to 1,000 g/m^2	Aerospace Industrial Sport and leisure
	Multiaxials	Strength and stiffness in multiple directions Control of fibre orientation ($0°$, $30°$ to $70°$, $90°$) Ability to optimise weight distribution in fabric Possible to mix fibres No crimp Less waste for complex lay-ups (cross plies) Reduced processing cost Heavy weights achievable	Aerospace Industrial

Table 3.2 Reinforcement forms

Aircraft primary structures are critical parts of the aircraft, e.g., spars, ribs, bulkheads. Reproduced by courtesy of Hexcel Composites Ltd., Cambridge, UK

3.4.2.2 Matrix

The role of the matrix is to support the fibres and bond them together in the composite material. It transfers any applied loads to the fibres, thereby keeping the fibres in their position and chosen orientation. The matrix also gives the composite environmental resistance and determines the maximum service temperature of a prepreg. When selecting prepregs, the maximum service temperature is one of the key selection criteria for choosing the best prepreg matrix. Matrix resins can be classified as:

Thermosetting resins

- Polyester and vinylester
- Epoxy
- Bismaleimide
- Polyimide
- Phenolic
- Others in development

Thermoplastic resins

- PPS
- PEEK
- PEI
- PAI
- Others

The most commonly used thermosetting resins used in the prepreg industry are epoxy, phenolic and bismaleimide. Their advantages and typical application areas are shown in **Table 3.3.**

3.4.3 Analysis of the process

As previously mentioned, the starting material for the autoclave moulding process is a prepreg that contains fibres in a partially cured (B-staged) resin. Typically, a prepreg contains approximately 42% by weight of resin. If this prepreg is allowed to cure without any resin loss, the cured laminate would contain about 50% by volume of fibres. Since nearly 10 weight percent of resin flows out during the moulding process, the actual fibre content in the cured laminate is about 60 volume percent, which is considered an industry standard for aerospace applications. The excess resin flowing out from the prepreg removes

Table 3.3 Matrix systems		
	Advantages	Applications
Epoxy	**Excellent mechanical performance** Good environmental resistance and high toughness Easy processing	**120 °C cure:** Aerospace, sport, leisure, marine, automotive, railways, transport **180 °C cure:** Aerospace, military
Phenolic	**Excellent fire resistance** Good temperature resistance Low smoke and toxic emissions Rapid cure Economic processing	Aerospace (interior components), marine, railways
Bismaleimide (and polyimide)	**Excellent resistance to high temperatures** Service temperature up to 260 °C Good mechanical characteristics Good resistance to chemical agents, fire and radiation	Aeroengines, high temperature components
Reproduced by courtesy of Hexcel Composites Ltd., Cambridge, UK		

the entrapped air and residual solvents, which in turn reduces the void content in the laminate. However, the recent trend is to employ a near net resin content, typically 34 weight percent, and to allow only 1–2 weight percent resin loss during moulding [7]. The trend towards using near net resin content means that you avoid the absorption of any amount of resin from the prepreg and you get what you put on the lay-up.

3.4.3.1 Lay-up process

The lay-up operation is performed in a clean room which is designated a certain 'class'. Each class indicates the maximum amount of particle permitted in a clean room, and also defines the temperature and humidity limitations. Most operations are carried out in a clean room classified as Class 100,000. However, some manufacturers perform their operations in a Class 400,000 environment, since it is easier to maintain.

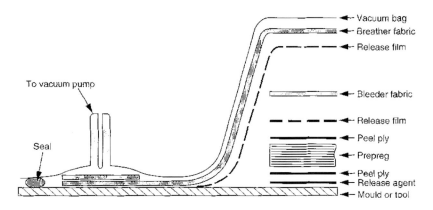

Figure 3.9 Vacuum bag lay-up

(Reproduced by courtesy of Hexcel Composites Ltd., Cambridge, UK)

Table 3.4 Typical vacuum bag materials	
Component	**Function**
Release agent	Allows release of the cured prepreg component from the tool.
Peel ply (optional)	Allows free passage of volatiles and excess matrix during the cure. Can be removed easily after the cure to provide a bondable or paintable surface.
Bleeder fabric	Usually made of glass fabric and absorbs the excess matrix. The matrix flow can be regulated by the quantity of bleeder, to produce composites of known fibre volume.
Release film	This layer prevents the further flow of matrix and can be slightly porous to allow the passage of only air and volatiles into the breather layer above.
Breather fabric	Provides the means to apply the vacuum and assists the removal of air and volatiles from the whole assembly. Thicker breathers are needed when high autoclave pressures are used.
Vacuum bag/ sealant tape	Provides a sealed bag to allow removal of air to form the vacuum bag. These are consumable materials and discarded each time, but some re-usable vacuum bag materials are available.

The prepreg material is removed from the freezer in this area. Plies are cut from the prepreg roll into the desired shape, size and orientation by a cutting device. The latter may be a mat knife or an automatic nesting (where plies of fabric are placed so that the yarns of one ply lie in valleys between the yarns of the adjacent ply) and cutting systems. Subsequently, the lay-up is performed using the auxiliary elements indicated in **Figure 3.9**. Prepreg plies are laid up in proper orientations determined through extensive tests, structural design and analysis studies. A new trend is to use laser projection systems during lay-up operations.

Figure 3.9 shows schematically all the components of a vacuum bag lay-up. This lay-up is ideal for high quality aerospace components. However, there may be some differences in the amount and type of vacuum bag materials depending on the part and material to be cured (shape, thickness, material properties, etc). Each layer in this figure has a role in the vacuum bag assembly as described in **Table 3.4** [8].

3.4.3.2 Curing process

After the vacuum bag is ready, the lay-up is inserted into the autoclave. A cure cycle, which has been developed through various engineering studies and tests, is then run (usually automatically), by a computer.

Initially, a vacuum is applied to the part to trap air, volatiles, and by-products. Heating begins at a specified heat up rate. As the prepreg is heated in the autoclave, the resin viscosity in the B-staged prepreg plies first decreases, then attains a minimum, and then increases rapidly (gels) as the curing (crosslinking) reaction begins and proceeds toward completion. **Figure 3.10** shows a typical two-stage cure cycle for a carbon fibre epoxy prepreg [7].

The first stage in this cure cycle consists of increasing the temperature up to 125 °C and dwelling at this temperature for nearly 60 minutes when the minimum resin viscosity is reached. During this period of temperature dwell, an external pressure is applied on the prepreg stack, causing the excess resin to flow out into the bleeder plies. A prepreg stack is a number of prepreg plies (layers) put or laid-up on each other to form a stack of prepreg material in the configuration defined by the part design regarding the fibre orientation. The resin flow is critical, since it allows the removal of entrapped air and volatiles from the prepreg and thus reduces the void content in the cured laminate. At the end of the temperature dwell, the autoclave temperature is increased to the actual curing temperature for the resin. The cure temperature and the pressure are maintained for 2 hours or more, until a pre-determined level of cure has occurred. At the end of the cure cycle, the temperature is slowly reduced while the laminate is still under pressure. The laminate is removed from the vacuum bag and, if needed, postcured at an elevated temperature in an air-circulated oven.

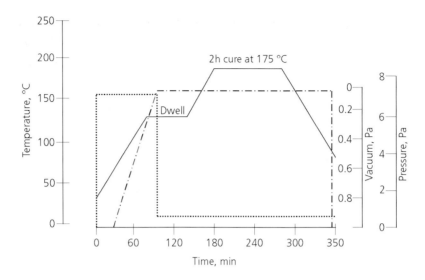

Figure 3.10 A typical two-stage cure cycle for a carbon fibre/epoxy prepreg

(Reprinted from P.K. Mallick, Fiber Reinforced Composite (Materials, Manufacturing and Design), 1988, p.332, by courtesy of Macel Dekker Incorporated)

The flow of excess resin from the prepregs is extremely important in reducing the void content in the cured laminate. In a bag moulding process for producing thin shell or plate structures, resin flow by face bleeding (normal to the top laminate face), is preferred over edge bleeding. Edge bleeding describes the flow of the resin through the edges of the part under heat and pressure. Face bleeding is more effective since the resin flow path before gelation is shorter in the thickness direction than in the edge directions. Since the resin flow path is relatively long in the edge directions, it is difficult to remove entrapped air and volatiles from the central areas of the laminate by the edge bleeding process.

The cure temperature and pressure are selected to meet the following requirements:

1. The resin is cured uniformly and attains a specified degree of cure in the shortest possible time.

2. The temperature at any position inside the prepreg does not exceed a prescribed limit during the cure.

3. The cure pressure is sufficiently high to squeeze out all of the excess resin from every ply before the resin gels (increases in viscosity) at any location inside the prepreg.

Loos and Springer [9] developed a theoretical model for the complex thermomechanical phenomenon that takes place in a vacuum bag moulding process. Based on their model and experimental works, the following observations can be made regarding the various moulding parameters:

1. The maximum temperature inside the lay-up. This depends on:

    - the maximum cure temperature

    - the heating rate and

    - the initial lay-up thickness.

The maximum cure temperature is usually prescribed by the prepreg manufacturer for the particular resin-catalyst system used in the prepreg and is determined from the time-temperature-viscosity characteristics of the resin-catalyst system. At low heating rates, the temperature distribution remains uniform within the lay-up. At high heating rates and increased lay-up thicknesses, the heat generated by the curing reaction is faster than the heat transferred to the mould surfaces and a temperature 'overshoot' occurs.

2. Resin flow in the lay-up. This depends on:

    - the maximum pressure

    - lay-up thickness and

    - pressure application rate.

A cure pressure sufficient to remove all excess resin from 16 and 32 lay-ups was found to be inadequate for removing resin from the layers closer to the bottom surface in a 64 ply lay-up. Similarly, if the heating rate is very high, the resin may start to gel before the excess resin is ejected from every ply in the lay-up.

Loos and Springer [9] have noted that the cure cycle recommended by prepreg manufacturers may not be adequate to remove excess resin from thick lay-ups. Since the compaction and resin flow progress inward from the top, the plies adjacent to the bottom mould surface may remain uncompacted and rich in resin, thereby creating weak interlaminar layers in the laminate.

Excess resin must be removed from every ply before the gel point is reached at any location in the prepreg. Therefore, the maximum cure pressure should be applied just before the resin viscosity in the top ply becomes sufficiently low for the resin flow to occur. If the pressure is applied too early, excess resin loss would occur, owing to very low viscosity in the pregel period. Conversely, if the cure pressure is applied after the gel time, the resin may not be able to flow into the bleeder cloth because of the high viscosity it quickly attains in the postgel period. Thus, the pressure application time is an important moulding parameter in a bag moulding process. In general, it decreases with increasing cure pressure, as well as increasing heating rate.

The uniformity of cure in the laminate requires a uniform temperature distribution in the laminate. The time needed for completing the desired degree of cure is reduced by increasing the cure temperature as well as increasing the heating rate.

Besides voids and improper cure, defects in autoclave-moulded laminates relate to the ply lay-up and trimming operations. Close control must be maintained over the fibre orientation in each ply, the stacking sequence, and the total number of plies in the stack. Since prepreg tapes are not as wide as the part itself, each layer may contain a number of identical plies laid side by side to cover the entire mould surface. A filament gap in a single layer should not exceed 0.76 mm, and the distance between two gaps should not be less than 38 mm. Care must also be taken to avoid filament crossovers. Broken filaments, foreign matter or debris should not be permitted. To prevent moisture pick-up, after removal of the prepreg roll from cold storage, it should be warmed to room temperature before use.

3.4.3.3 Cure cycle development and quality control

Since the cure of composites is important for the desired properties, industry has taken a very conservative approach toward the parameters, and the cure of these materials. Appropriate tests are conducted to ensure that a particular piece of pre-impregnated composite material yields acceptable parts. Traditional tests involve making a panel and performing various mechanical tests, such as tensile, flexure, short beam shear, and so forth, as well as resin content, flow and volatile content. Chemical tests, such as high pressure liquid chromatography (HPLC), thermogravimetric analysis (TGA), differential scanning calorimetry (DSC), rheology, infrared (IR) spectroscopy, etc., are also performed for material characterisation, development of process parameters, quality control, shelf life renewal, etc. These tests show the chemical and physical state of the uncured resin. Cured resin systems are also studied by using TGA, thermomechanical analysis (TMA), dynamic mechanical analysis (DMA), etc. Tests can indicate:

- the chemical advancement of the resin system

- whether the resin is at the proper stage for cure

- the ratio of reactants to reaction products

- at which temperature the volatiles will evolve

- the viscosity and
- whether the supplier has sent the correct material.

After the parts are cured, various Non Destructive Inspection (NDI) techniques are involved for quality assurance purposes. For composite parts ultrasonic inspection methods like through transmission and pulse echo are mostly used. Some thermographic and tomographic methods are also applied for contoured, complex and structurally-critical parts like helicopter blades. X-ray methods are applied mainly for sandwich structures to find the defects in the core region.

3.4.3.4 Tooling

The primary purpose of tooling is to provide the contour configuration that the part assumes when it is finally cured. Transferring heat to the part lay-up is another function of the tool. Tools are generally made from the durable materials listed in **Table 3.4**, although single parts can be made on a plaster tool that is usually destroyed by the heat cycle.

Table 3.5 Properties of typical composite tooling materials			
Material	CTE 10^{-6}/K	Thermal conductivity w/m.K	Approximate fabricated cost \$/m^2
Fibreglass-epoxy	7.9	-	175–278
Graphite-epoxy	-0.9	0.022	220–320
Aluminium	23.0	0.221	390–600
Steel	13.9	0.048	250–450
Electroless nickel	13.3	0.035	370–580

It is highly desirable to have the CTE of the tool closely match the thermal expansion of the part being made. This is the basis of the trend towards using steel or high temperature carbon-epoxy tools for composites. For small parts, particularly if they are flat, aluminium or steel plates are frequently used as the tool material. For larger, more complex parts, electroless nickel-plated tools or high-temperature epoxy with carbon reinforcement is used. The ability of electroless nickel to provide a uniform deposit, even into deep recesses, ensures that the finish on the mould will duplicate the original surface. The natural lubricity of the coating provides smooth flow during injection and quick and easy release of the part from the mould. On pre-polished cavities, electroless nickel is self-polishing and helps to maintain the smoothness of the mould. Because of its high hardness, even at elevated temperatures, electroless nickel minimises erosion and abrasion of mould parts and helps to extend their lives. It provides excellent protection against the corrosive fumes produced during moulding such plastics as ABS, PVC and materials with vinyl additives.

Tools may also contain special inserts, stops, or guides to facilitate the part construction. Built-in vacuum and static ports, as well as permanent thermocouples have also been built into tools. Static parts (vacuum transducers) are used to measure the vacuum applied to the part (vacuum bag). These features are high maintenance items and tend to increase tooling cost, which must be offset by the gain in efficiency over the long-term use of the tools for several cure cycles. Tooling substructure has a significant impact on the air flow under the tool and therefore on the heat transfer into the part during the cure cycle. The current trend is to incorporate large holes in the tool substructure and to align the substructure supports parallel to the air flow to improve heat transfer.

Tooling is generally heated in the autoclave with convective air flow; ovens are rarely used, because a higher pressure is usually necessary for part consolidation. Electrically-heated tools have been used infrequently, because the cost of such systems is high and there is a potential shock hazard [10].

3.4.4 Future development of autoclave moulding process models

As new materials with smaller processing windows (and less room for error) become available, the use of a computer process model to predict the behaviour of the composite material during cycle, based on its constituent properties, would reduce fabrication costs and ensure a better product quality.

On-line control applications using process models allow corrections to the cycle. However, these applications need to be able to recognise deviations in the initial conditions, as well as deficiencies in the desired processing conditions. For this to occur, on-line control applications require two additional components:

- in situ measurements (sensor feedback) of critical variables and

- a mathematical definition of acceptable quality as a function of time.

An expert system collects and combines the information from all of the defined approaches to optimise and control the process on-line. An expert system includes three primary components: a representation of experience (which includes predictions made using a process model), a decision making shell, and a closed loop control capability. The expert system approach can be viewed as a new application of process modelling. While the conventional process model remains stagnant, the use of intelligent systems grows as understanding of the process increases.

Two examples of commonly used expert control systems are di-electric and ultrasonic cure monitoring. Presently, ultrasonic sensing is limited to thin laminates. Using these two systems, the state of the resin is monitored during the cure cycle. This enables optimisation of the cycle by the correction application of pressure, heating and cooling. This ensures a closed loop control (automatic cure cycle generation) when suitable software is employed. In other words, there is a 'closed loop' where the data from the sensors are evaluated continuously and the cure cycle is always modified and optimised automatically by computer. 'Automatic cure cycle generation' is generation of the cure cycle (heat-up, cool-down rate, dwelling times, etc.) by computer through evaluating the sensor data. No specific cure cycle program is entered before beginning the cure as in conventional autoclave or oven moulding procedures.

Hence, it is possible to generate optimum cure cycles that are specific for each part, and achieve on-line quality control without the need for laborious studies. Presently, these systems are used in Research and Development studies and prototype development. However, as their use increases, coupled with developments in sensing systems and software, this system could become a common tool for manufacturing.

3.4.5 Applications

Composite structural elements are now used in a variety of components for automotive, aerospace, marine, and architectural structures, in addition to consumer products such as skis, golf clubs, and tennis rackets. Much of the current composite technology has evolved from aerospace applications. Composite technology is initially developed for aerospace applications but, as material and processes costs decease, it finds a niche in other industries.

Composites have high specific strength and stiffness, characteristics which make them ideal for use in air vehicles, where performance and manoeuvrability are highly weight-dependant.

The use of composite structures is increasing annually. The percentage of composite structures on new civilian and military aircraft increases to approximately 90% in some applications.

Figure 3.11 Almost all composite MD-902

(Photograph reproduced by courtesy of MD HELICOPTERS, Inc., Mesa, Arizona, USA)

Figure 3.11 depicts a helicopter which is almost entirely composed of composite materials. It is an example of the use of composite materials and autoclave mouldings in air vehicles. Autoclave moulding is the premier manufacturing method in the aerospace industry, due to its quality and performance. However, manufacturing methods such as RTM are being developed as an alternative to autoclave moulding. The latter is also used in the manufacture of yachts and formula cars.

3.4.6 Advantages and disadvantages

Autoclave moulding is the premier method for the manufacture of quality composite parts for the aerospace industry. Also good dimensional stability is obtained and laminates of variable thickness (as well as contours of varying complexity) can be produced.

The main disadvantage is cost, since:

- pressurisation, controlled heating and cooling, vacuum is expensive
- the operation is very time-consuming (long cycle times)
- the operation is unsuitable for serial production (labour intensive)
- the material is very expensive
- the operation requires auxiliary materials, since vacuum bags are single-use items

However, since quality and performance are the main criteria for both civilian and military applications, the expense of manufacture is offset through lower operating costs.

Acknowledgment

With thanks to Hexcel Composites Ltd., for allowing us to use extracts from their Prepreg Technology Manual [8], available from the company on request.

References

1. M. P. Groover, *Fundamentals of Modern Manufacturing*, Wiley, New Jersey, USA, 1996.

2. *Plastics Handbook*, Modern Plastics Magazine, McGraw-Hill, New York, USA, 1994.

3. M. M. Schwartz, *Composite Materials, Volume 2: Processing, Fabrication and Applications*, 1st Edition, Prentice Hall, New Jersey, USA, 1997.

4. M. M. Schwartz, *Composite Materials Handbook*, McGraw-Hill, New York, USA, 1984.

5. *Engineered Materials Handbook, Volume 1: Composites*, ASM International, Ohio, 1988.

6. George Lubin in *Handbook of Composites*, Van Nostrand Reinhold Company Inc., New York, USA, 1982.

7. P. K. Mallick, *Fibre Reinforced Composites: Materials, Manufacturing, and Design*, 1st Edition, Marcel Dekker, New York, USA, 1988, 331.

8. *Prepreg Technology*, CIBA Composites Publication No. FGU 265, 1995.

9. A. C. Loos and G. S. Springer, *Journal of Composite Materials*, 1983, **17**, 2, 135.

10. R. R. Sanders and S. Taha in *Engineered Materials Handbook*, Volume 1, ASM International, Ohio, 1988, 702.

4 Closed Mould (Matched Die) Processes

S. Kenig

4.1 Introduction

Closed mould or matched-die processes are widely used in the composites industry for the fabrication and manufacturing of three-dimensional components and products.

Among the various techniques used in closed mould processes are:

a) Transfer moulding - in which a resin is transferred from a reservoir into a three-dimensional closed heated mould, with curing taking place due to the temperature of the resin and the mould. By this technique pre-placed layers of fabrics in the closed mould are processed where, in a subsequent stage, resin is introduced into the mould by pressure and/or vacuum.

b) Compression moulding - where a pre-weighted amount of premixed compound is placed in a heated three-dimensional mould. As the mould closes, the material flows and fills the mould cavity. Simultaneously with the flow stage, curing takes place and it continues while pressure is applied when the mould is in a closed position. SMC, BMC and MC are the material systems that are processed by this moulding method.

c) Injection moulding - in this process, a thermoplastic or a thermoset system is injected under high pressure into a closed three-dimensional cavity, which is held at low or high temperature, respectively. Cooling and solidification of the thermoplastic-based system or curing of the thermosetting system take place while the material is under high pressure.

This chapter will survey the various fabrication methods used in closed mould processes comprising the material systems and associated process conditions, the resultant properties as affected by the process parameters and materials composition. In addition, modelling of the various closed mould processes will be summarised and analysed.

4.2 Transfer moulding

4.2.1 Introduction

Transfer moulding fabrication methods include RTM. RTM has variants, such as Assisted RTM (ARTM) and Thermal Expansion RTM (TERTM). All of these methods have been recognised as cost-effective techniques for the fabrication of high performance polymer composite components. RTM is a reactive processing method in which a liquid resin, with its appropriate curing agents, flows into a mould and cures in the mould containing fabric reinforcement. As the pressure used to fill and force the liquid to flow is low (690 kPa) the mould construction and clamping forces are simple and low, respectively.

The driving force for RTM was the automotive industry, consumer products and, more recently, marine and aviation markets. The elimination of emissions like styrene, by using closed moulds in wet systems for processing, has contributed to the expansion of RTM in the composite industry.

This chapter will review the main materials compositions used and the basic issues associated with the RTM process. The latter include the wetting and flow of the liquid resin within the reinforcement, as well as the properties and applications of the end product.

4.2.2 Materials and processes

RTM is an ideal technology for the fabrication of high strength/high rigidity parts in medium series production (tens of thousands). Since the reinforcement is placed in the mould prior to resin injection, improved preform technology is the key issue for developing this fabrication technique to large scale production [1]. Furthermore, as the resin flows through the dry reinforcement, viscosity is the main parameter in the filling of the cavity. Typical viscosities are in the range of 0.1–1 Pa-s with pot life (the length of time that a catalysed thermosetting resin system retains a viscosity low enough to be used in processing), between 120 to 1000 seconds.

As the resins possess low viscosities, moulding pressures are in the range of 90 to 600 kPa [2]. Consequently, the tonnage requirement is low (20 tons/m^2) and the cost of equipment and tooling is in the range of tens of thousands of dollars.

The resins used are mainly epoxy- and polyester-based, and for high service temperatures (180 °C), bismaleimide- and phenolic-based resins are used.

RTM sales are increasing at the expense of autoclaving fabrication in aerospace and sports market segments. Typical properties of RTM products are as summarised in **Table 4.1** [2].

Table 4.1 Typical properties of RTM products	
Tensile modulus	15 GPa
Tensile strength	270 MPa
Flexural modulus	24 GPa
Flexural strength	550 MPa
Notched izod	250 kJ/m^2
Barcol hardness	40

Recent developments in RTM are presented for fast cured vinyl-ester formulations, which provide short cycle time, low viscosities, good chemical stability and high toughness [3]. These epoxy-based backbone polyesters with reactive vinyl end groups have been specifically formulated for RTM processes using 50 to 2,500 kPa injection pressure, when keeping the mould at 100 °C at a glass content (chopped mat) of 53%. The resulting properties are:

- a tensile modulus of 14 GPa

- a tensile strength of 105 MPa

- a flexural modulus of 11 GPa and

- a flexural strength of 350 MPa.

4.2.3 Kinetics of resin curing

Understanding closed mould processes like compression moulding, transfer moulding and injection moulding requires an in-depth knowledge of the kinetics governing the resin cure reaction. The heat of reaction and the rate of heat generation are the two decisive parameters that effect the curing kinetics. The final degree of cure and the thermal history determine the physical, mechanical and thermal properties of the cured material.

The basic kinetics (isothermal) is described by Kamal and Sourcow [4].

$$d\alpha/dt = K_1 + K_2 \, \alpha^m \, (1-\alpha)^n \tag{4.1}$$

Where:

α is the degree of curing conversion

$d\alpha/dt$ is the rate of curing

K_1 and K_2 are temperature-dependent constants

n and m are kinetic constants independent of temperature.

Various parameters have been experimentally evaluated for a variety of resin systems, e.g., epoxies [5, 6], unsaturated polyesters [7] and polyurethanes [8].

4.2.4 Rheological analysis

In flowing reactive systems, the rheological behaviour is of fundamental importance for the analysis of the filling and flowing stage of the various fabrication methods: compression moulding, transfer moulding and injection moulding. Initially, the viscosity of the resin system is low, due to the combination of low molecular weight and elevated temperature. As the molecular weight increases with time, the viscosity increases to the point where flow terminates. Roller [9, 10] reviewed the models for analysis of the viscosity variations with time and temperature in reactive-thermosetting systems.

The practical model assumes a first order reaction where the viscosity varies with time, $\eta_{(t)}$, according to the following equation:

$$\eta_{(t)} = \eta_o \exp (K \, t) \tag{4.2}$$

Where:

η_o is the initial viscosity

K is the reaction rate constant

t is the reaction time.

Assuming that η_o varies with temperature, T, then according to Arrhenius:

$$\eta_o = \eta_\infty \exp (E_\eta /RT) \tag{4.3}$$

and

$$K = K_o \exp (-E_\alpha /RT) \tag{4.4}$$

Where:

η is the viscosity constant

K_o is a reaction constant

E_η is the activation energy for flow

E_α is the activation energy for curing

4.2.5 Resin flow through the fabric

The mould-filling stage in RTM or Structural Reaction Injection Moulding (SRIM), in which a fabric is placed in the mould, is a complex procedure. This is due to the special characteristics of resin flow through the fabric bundles. Darcy's Law is the basic model used to describe the flow of the resin through the fabric. It assumes that the resin follows Newtonian behaviour and the flow through the fabric could be described by a flow in a porous media. Accordingly, the flow rate (Q) is related to the pressure gradient (P/L) as follows:

$$Q = S \, (A/\mu) \, (\Delta P/L) \tag{4.5}$$

Where:

S is Darcy's constant for permeability (1 Darcy = $0.987 10^{-2}$ m²/s)

μ is the Newtonian viscosity

A is the cross-sectional area.

The permeability constant (S) was modelled by Kozery-Carmen as a bundle of network capillaries. Consequently, in the case of flow in a bundle network, S is given by:

$$S = \phi^3/[CA_v^2 \, (1 - \phi)^2] \tag{4.6}$$

Where:

ϕ is the porosity

A_v is the surface area/volume

C is a constant.

For flow along the fibre direction, permeability is described by:

$$S_x = r_f^2 \, (1-V_f)^3/(4s_x V_f^2) \tag{4.7}$$

Where:

S_x is the permeability in fibre direction

r_f is the fibre radius

s_x is the Kozery constant and

V_f is the fibre volume fraction.

For flow perpendicular to fibre direction, the permeability constant (S_z) is given by:

$$S_z = r_f^2[(V_a^1/V_f)^{1/2}-1]^3/[4s_z(V_a/V_f-1)] \tag{4.8}$$

Where:

V_a^1 is the critical fibre flow volume fraction (0.80–0.85)

S_z is the Kozery constant.

The models discussed above have been used in computer-aided design packages for process optimisation.

4.3 Compression moulding

4.3.1 Introduction

Premix (SMC, BMC) and MC involve the incorporation of resin and curing agent, filler and reinforcement, charging the compound system into a three-dimensional shaped cavity, which is then placed in a compression press. The application of pressure and elevated temperature causes the compound to flow in the mould. The flow pattern in the mould results in the orientation of the fillers and fibres. As the compound flows and, following the complete filling of the heated mould, curing takes place. The moulded products are ejected from the mould when the crosslinking level has reached a sufficiently high level. Post curing is carried out following ejection of the shaped article in a heated oven, in cases where a maximum level of curing is needed. The resultant, compression-moulded parts possess a spectrum of properties, including high rigidity and strength (tensile, compression, impact) and good surface properties (gloss, smoothness, paintability).

In principle, the thickness of compression-moulded compounds are not limited (in contrast to injection moulding).

Section 4.3.2 will review the materials composition, followed by the process and models used, the end-use properties and applications of compression moulding.

4.3.2 Materials

A typical compound composition used in BMC and SMC includes base resins, catalysts and accelerators, fillers and chopped glass fibres, thickeners and additives.

4.3.2.1 Resins

Typically, the resins used for BMC, SMC and MC are polyester-based: orthophthalic, isophthalic, with styrene or acrylic monomers (crosslinkers). In some cases, vinyl toluene and di-allyl-phthalate (DAP) are used as monomers. For general use, styrene-type monomers are used. Whereas, for low shrink characteristics, acrylic types are used. Also, for high hot strength (heat deflection temperature), vinyl type monomers are employed.

4.3.2.2 Catalysts and accelerators

For polyester-type resins, peroxides are the main catalysts. Common examples are benzoyl peroxide (BPO) and t-butyl perbenzoate, which is a high temperature catalyst.

4.3.2.3 Fillers

Fillers are used in BMC, SMC and MC as low cost inert additives. In addition, fillers can:

- modify the viscosity and act as flow control agents
- reduce the CTE
- serve as heat sink for exotherms that develop during the curing reaction
- increase hardness, rigidity and dimensional stability.

The most common powder filler is calcium carbonate. Ground clay (kaolin clay) is also used, however it causes more discolouration and higher shrinkage than calcium carbonate. Talc is often used for improved electrical strength and resistance to humidity. Aluminium hydrate is commonly used for fire retardancy as it contains 35% of hydration water, which is released upon exposure to fire. Powdered polyethylene is used in low profile (low shrinkage) formulations, since it improves surface quality and impact strength. Antimony trioxide and chlorinated waxes are also used for self-extinguishing properties.

4.3.2.4 Glass fibres

Chopped E glass type fibres are used as the reinforcing agent. The fibres are coated with organic binders, with lengths of 0.3 cm, 0.6 cm, 1.3 cm, and up to 5.1 cm. Occasionally, chopped fabrics are used as an additional reinforcement.

4.3.2.5 Thickeners

To ensure controlled mouldability, thickeners are used for controlled increase of the viscosity of the polyester resin. Typical thickeners are magnesium oxide and hydroxide, calcium oxides and hydroxides.

4.3.2.6 Miscellaneous additives

In addition to the additives mentioned in the previous section, stearates are used for internal mould release and as various pigments for colouring.

4.3.2.7 Formulations

A variety of formulations are available for MC, BMC and SMC systems. The most common formulations are based on polyester resins. The formulations are categorised as follows:

a) general purpose,

b) low shrink,

c) low shrink and thickened.

Each category is formulated for the BMC and SMC systems.

Table 4.2 summarises the compositions of typical BMC systems.

As shown in **Table 4.2**, the usual ratio of resin to fibre content is 2:1.

Table 4.3 summarises the compositions of a typical SMC system.

As shown in **Table 4.3**, the ratio of resin to the reinforcement used for SMC's is usually higher than that used for BMC.

Table 4.2 Typical BMC formulations (% by weight)			
	General Purpose	Low Shrink	Low Shrink and Thickeners
Resin	30	19	18
Low shrink resin	—	13	7
Low shrink modifier	—	—	5
Styrene	6.4	5.4	4.7
Catalyst (BPO)	0.6	0.3	0.3
Zn stearate	1.0	0.3	1.0
$CaCO_3$ and clay	47	47	48
0.6 cm glass fibres	15	15	15
Thickeners	0	0	1

Table 4.3 Typical SMC formulations (% by weight)		
	General Purpose	Low Shrink
Resin	44	30
Low shrink resin	—	15
Styrene	5	5
Catalyst (BPO)	0.4	0.7
$CaCO_3$ and filler	15.5	18.6
Mould release	0.1	0.7
Glass mat	35	30
Thickeners	—	1.0

4.3.2.8 Properties of BMC and SMC systems

Various formulations yield a spectrum of physical and mechanical properties of the final products. **Table 4.4** and **Table 4.5** shows typical properties of BMC and SMC formulations, respectively.

Table 4.4 Typical properties of BMC	
Specific gravity	1.4–2.0
Flex modulus (GPa)	10–18
Flex strength (MPa)	40–180
Tensile modulus (GPa)	10–15
Tensile strength (MPa)	20–70
Impact (notched izod)	2–8
HDT (°C)	150–200
Barcol hardness	35–70

Table 4.5 Typical properties of SMC			
	General Purpose	High Toughness	Low Shrink
Specific gravity	1.14	1.17	1.67
Flex modulus (GPa)	4.2	3.9	11.6
Flex strength (MPa)	95	150	85
Tensile strength (MPa)	70	76	35
Tensile elongation (%)	1.8	4.0	—
Barcol hardness	50	45	60

High modulus SMC (High Modulus Compound or HMC) that contain 45% glass fibres have been developed for special applications in the automotive industry [11]. Their typical properties are:

- modulus of 15 GPa
- tensile strength of 160 MPa

- flexural strength of 300 MPa

- unnotched Charpy of 115 J/cm^2

Low density SMC have been manufactured using low pressure processing [12], to meet the demands of the automotive industry. Low density is achieved by adding microspheres, yielding densities in the range of 1.3 to 1.5 x 10^{-3} kg/m^3 (standard densities 1.8–1.9 x 10^3 kg/m^3). Low pressure compression moulding results from easy flow formulations that require pressures of 15 to 35 MPa. The resulting properties are:

- flexural modulus of 8 GPa

- flexural strength 170 MPa

- fibre content of 30%

Vinyl-ester SMC formulations have also been developed especially for the automotive industry. Products such as large exterior body panels and under the hood components have been used successfully due to their high corrosion resistance and high temperature rating. Typical formulations contain: 30% glass fibres, thickeners and 35% fillers (calcium carbonate, clay, and aluminium hydrate [13, 14]). The resulting properties are:

- flexural modulus of 14 GPa

- flexural strength of 210 MPa

- T_g of 170–200 °C

4.3.3 Modelling compression moulding

In the compression moulding process, a thermosetting-based compound is initially placed in a heated cavity. As force is applied, the two halves of the mould begin to close and force the compound to flow into the empty parts of the cavity. This type of flow is often called 'squeezing flow'. As the polymer flows into the heated cavity, it crosslinks.

Modelling of the compression moulding process comprises the 'squeezing flow' phenomenon and curing kinetics. In squeezing flow, a compounded reactive resin flows radially as a result of the compressive load which develops as the mould is closed. In addition the gap (distance) between the closing mould walls changes with time. The problem was analysed by Leider and Bird [15, 16], who assumed that the system exhibited shear flow. However, squeezing flow has an additional important feature: elongational flow in the hoop as well as in the flow direction. The increasing radii that take place as the compound flows outward results in increased cross-sectional area and deceleration of flow. As a result, the fibres tend to orient perpendicular to the radius at each point.

An additional orientation force results due to the 'fountain-flow' phenomenon that gives rise to orientation in the radial direction in the skin layers (the relatively dense material that may form the surface of a cellular plastic). Consequently, short-fibre containing compounds will have characteristic fibre orientation in the inner layers (transverse orientation) and skin layers (flow direction orientation).

The problems of a crosslinking reaction which involves heat transfer has been investigated by Broyer et al. [17]. Accordingly, in the case of BMC and MC, where discontinuous fibres (chopped fibres) are used, movement and flow of the fibres occurs. This results in preferred fibre orientation—as discussed previously—due to the elongation and shear stresses that occur in squeezing flow.

In the case of SMC (where continuous fibre reinforcement is present), the unpolymerised resin flows upon squeezing. This permits the woven fibres to bend and move to a certain degree, due to the stresses induced by the flowing resin and the closing mould.

Computer simulations which describe the transformation of a flat sheet into a three-dimensional shaped product are available. This transformation involves the movement of resin and fibres. For woven fabrics, simulation of the transformation was first carried out by Van der Ween [18] using 'fishnet analysis'. For unidirectional fibres, mapping of the shaped product was performed by Gutowski [19]. A more detailed approach was developed using the continuum mechanics approach. Accordingly, the shaping process is carried out gradually [20], as the stresses applied result in the corresponding deformations. This approach requires the relevant constitutive equations for anisotropic systems.

Numerical analysis using finite element representation schemes are used to simulate the complete compression moulding process and especially the fibre orientation [21, 22]. The resulting mechanical properties can be calculated using Halpin-Tsai equations [23].

4.4 Injection moulding

4.4.1 Introduction

Injection moulding of fibre or filler reinforced polymer compounds is a widely used technique for fabrication of three-dimensional composite components. Though the resin systems used are both thermoplastic and thermoset in nature, this section will deal with reactive systems only. The main advantage of injection moulding is the high production rate and the high automation level of the process. In injection moulding, the filling of the mould is carried out under high pressures, resulting in short filling times (seconds). Curing takes place in the heated mould under high pressure (10–20 MPa). A few variations of the injection moulding process are practiced in the industry. One example is the injection moulding of pre-compounded resin, short fibres and/or fillers. Another variant is SRIM, which involves the injection moulding of resin under high pressure into a heated mould, where fabric reinforcement occurs. This section will review the materials used and their final properties, as well as discussing flow analysis and end-use applications.

4.4.2 Materials and processes

Two main fabrication methods based on reactive systems containing fibre reinforcement. They both involve the injection moulding process, namely: (a) injection moulding of MC and (b) SRIM. The first fabrication method is based on polyester (BMC/SMC) [24], whereas phenolics are mainly based on novolac (a linear thermoplastic B-staged resin) systems [25].

For SMC injection moulded parts containing 15% glass fibres, typical rheological values are:

- flexural modulus of 8.5 GPa

- flexural strength of 110 MPa

- izod impact (notched) of 3 ft.lb/inch

The second fabrication methods involves the placement of fabric or mat reinforcement into the cavity, followed by polyurethane (mixed polyol and isocyanate) injection. As is the case in the RTM process, flow of the resin through the fabric and wetting of the fibres are the main variables. Pressures in the range of 10 to 20 MPa are used, resulting in a high tonnage press system (100 tons/m^2). Viscosities of resin components is in the range of 0.1 to 1 Pa-s (40–60 °C temperature) with start cream times (2–20 seconds) [26]. The cream time is a measure of time from the start of hardening (the change from liquid to gel). This system classified as a fast reaction, whereby the components are mixed by direct impingement under high pressure. This fabrication method is particularly popular in the automotive industry where bumper beams, doors and chassis components are produced [27]. The polyurethane system has the potential of being low cost, in addition to its low specific weight (27). Furthermore, mineral fillers can be added, e.g., Reinforced Reaction Injection Moulding (RRIM).

4.4.3 Fibre orientation in injection moulding of thermosets

Short fibre reinforced thermosets fabricated by injection moulding offer some of the advantages of continuous fibre composites combined with the economical aspects of the injection moulding process. The user and designer of short fibre composites are usually confronted with the discrepancy between the mechanical property data (as determined by the standard testing methods) and the actual properties exhibited in the final moulded article. These discrepancies result from the anisotropy of fibre distribution. Since the fibre orientation distribution is critical to the mechanical performance in short fibre composites, a great deal of effort has been made to develop numerical methods for determining the local fibre orientation as a function of the injection moulding processing condition.

A variety of software packages are available for calculating the state of orientation resulting from flow combined with crosslinking. The origin of orientation in short fibre injection moulding was summarised by Kenig [29]. Basically, orientation is highly affected by the elongational type flows that prevail in the injection moulding of fibre-filled compounds. In the flow front, the 'fountain flow' is responsible for the fibre orientation in the skin layers in the flow direction, while the radial spreading flow in the vicinity of the gate gives rise to transverse or random orientation in the core regions. Shear-induced orientation takes place at a certain distance from the cavity wall. This is due to the high viscosity of the flowing compounds, which is caused by a higher state of curing in this region (due to a high mould temperature).

The combination of the flow and thermal histories can explain the complex orientation patterns in injection moulding of fibre containing thermosets. Usually a layered structure is formed.

References

1. P. Mapleston, *Modern Plastics International*, 1989, **19**, 11, 48.

2. R. Alaka, S. Ide and T. Kobayashi, *Reinforced Plastics*, 1998, **42**, 2, 26.

3. D. Babbington, J. Barron, M. Cox and J. Enos, Presented at the 42nd SPI Annual Conference and Expo '97, Cincinnati, OH, USA, 1987, 23-D.

4. M.R. Kamal and S. Sourour, *Polymer Engineering and Science*, 1973, **13**, 1, 59.

5. W.I. Lee, A-C. Loos and G.S. Springer, *Journal of Composite Materials*, 1982, **16**, 6, 510.

6. W.M. Sanford and R.L. McCullough, *Journal of Polymer Science Part B: Polymer Physics*, 1990, **28**, 7, 973.

7. J.F. Stevenson, *Polymer Engineering and Science*, 1986, **26**, 11, 746.

8. L.J. Lee and C.W. Macosko, *International Journal of Heat Mass Transfer*, 1979, **23**, 1479.

9. M.B. Roller, *Polymer Engineering and Science*, 1975, **15**, 5, 406.

10. M.B. Roller, *Polymer Engineering and Science*, 1986, **26**, 6, 432.

11. *Modern Plastics International*, 1995, **25**, 3, 78.

12. J. de Gaspari, *Plastics Technology*, 1996, **42**, 3, 32.

13. M. Winkler, Presented at the 45th SPI Annual Conference, Washington, DC, USA, 1990, 1E.

14. U.L. Bürgel, *Kunststoffe German Plastics*, 1990, **80**, 11, 29.

15. P.J. Leider and R.B. Bird, *Industrial and Engineering Chemistry Fundamentals*, 1974, **13**, 336.

16. P.J. Leider, *Industrial and Engineering Chemistry Fundamentals*, 1974, **13**, 342.

17. E. Broyer and C.W. Macosko, *American Institute Chemical Engineering Journal*, 1976, **22**, 268.

18. F. Van der Ween, *International Journal for Numerical Methods in Engineering*, 1991, **31**, 1415.

19. T.G. Gutowski, P. Hoult, G. Dillan and J. Gonzalez-Zugasti, *Composites Manufacturing*, 1991, **2**, 3/4, 147.

20. C.L. Tucker and R.B. Pessenberger in *Flow and Rheology in Polymer Composites Manufacturing*, Ed., S.G. Advani, Elsevier, Amsterdam, 1994, 257.

21. L.G. Reifsneider and H.U. Akay, Presented at ANTEC 92, Detroit, MI, 1992, 1695.

22. M.R. Barone and D.A. Chaulk, *Journal of Applied Mechanics*, 1986, **53**, 361.

23. J.C. Halpim and N.J. Pagano, *Journal of Composite Materials*, 1969, **3**, (4) 720.

24. R.S. Drake and A.R. Siebert, Presented at the 42nd SPI Annual Conference and Expo '87, Cincinnati, OH, USA, 1987, 23-D/1.

25. D. Brosius, *Modern Plastics*, 1996, 73, 11, B62.

26. G. Graff, *Modern Plastics International*, 1998, **28**, 8, 68.

27. S. Kenig, *Polymer Composites*, 1986, **7**, 1, 50.

5 Filament Winding

L. Parnas and S. Ardıç

5.1 Introduction

Filament winding is a relatively simple process in which a band of continuous reinforcements (fibres) or mono-filaments is wrapped around a rotating mandrel and cured to produce closed-form hollow parts. The winding operation is achieved by the use of specially designed machines. These machines, which can have up to 6-axes (even 7-axes with robotic enhancements), allow the operator to control various parameters, including the winding speed, winding angles, fibre placement, resin temperature and fibre tension. The filaments (or reinforcement bands) are wrapped around the mandrel as adjacent bands of repeating patterns. These cover the mandrel surface to produce one complete layer. The winding is continued with successive layers, which can be in different winding angles until the number of layers required by the designer is reached. The winding angle may vary with respect to the mandrel axis. Winding angles which are very close to 90° are termed hoop winding, whereas winding using other angles is termed helical winding.

If the fibres are passing through a resin bath before the winding on the mandrel, the process is called wet winding. If prepreg fibres are used, the process is termed dry winding. Rarely, the reinforcements are wound on the mandrel without any resin application and the reinforcement on the mandrel is impregnated with resin later: a process known as post-impregnation.

Wet winding is the most commonly used filament winding operation. Generally, curing is performed at elevated temperatures with no pressure application for wet winding. However, an autoclave curing process may be preferred for the dry winding procedure. The production process is completed with the removal of the mandrel. Machining operations may be performed for a better surface finish as well as the use of surface mats or veils at the end of winding process. Indeed, by using sophisticated modern filament winding machines, obtaining a product that needs no surface finishing operation is possible.

The main advantages of the filament winding technique are:

- Filament winding has a high repeatability. A part with the same properties can be produced repeatedly as well as the layers in the product.

- Continuous fibres are used in the loading direction: directional strength can be easily satisfied by changing the winding angle and winding pattern.

- Capital cost and process cost are low. For example, an autoclave is not required.

- Huge parts can be manufactured.

- In relation to prepreg materials, material cost is low.

- High fibre volume percentages are used, thus, high strength values can be obtained.

The disadvantages of filament winding can be summarised as:

- Complicated shapes require very complex mandrel designs and extra cost.

- Reverse curvature parts cannot be produced.

- Mandrels are indispensable and are expensive tools.

- The surface quality obtained is poorer than that obtained by autoclave production, and generally surface machining is required.

There are several parameters involved in the basic filament winding process. Some of these are structural aspects, design requirements, materials and equipment/tools, etc. Structures produced by using the filament winding technique are mostly surfaces of revolution which are obtained by rotating a plane curve around an axis. Therefore, their cross-sections are generally circular. In addition to these axisymmetric parts, asymmetric ones can also be produced. Filament winding has many applications:

- shafts: aeroplane, helicopter, windmill blades.

- liquid petroleum gas (LPG), natural gas, oxygen tanks.

- pipelines, pressure vessels, storage tanks, rocket motor cases, rocket launch tubes, structural elements, electrical insulator tubes, recreational equipment.

As these examples show, the products may be plain cylinders, pipes or tubes. The most common products vary from a few centimetres to metres in diameter. Pressure vessels and tanks (where the technique is satisfactorily used) are generally composed of a cylinder whose ends are closed by domes or end closures. These domes can be integrated either before or after winding. Instead of bands of filaments, tapes can be wound for large products in order to shorten the winding time.

The structures that are produced by using the filament winding technique are designed for various loading conditions. These include internal or external pressure, axial loads, torsion or bending. Depending on the designer's requirements, a combination of helical and hoop layers can be used. The material selection is also an important step. Glass, carbon and aramid fibres are used as reinforcements. Thermoset (and on rare occasions, thermoplastic) resin systems are used to bind the reinforcements. The fibre coatings and winding speed are crucial factors affecting the fibre wet out and, as a result, the performance of the product.

The winding machines were initially simple lathe type or chain-driven machines. These kind of machines are still widely used. Contemporary winding machines are all numerically-controlled 3- to 6-axis machines. Using these developed machines, it is possible to produce shapes, such as spheres, elbows or 'T'-shaped connection elements and irregularly-shaped parts. Of course, an important element of the filament winding process is down to the personnel operating the machinery. Hence, operators should be aware of the importance of all the parameters involved in the process and their effects on the performance of the final product. The use of complicated winding machines requires well-trained operating staff.

The materials, process, machines and tools needed for filament winding will be discussed in this chapter. Since this handbook is concerned with the fabrication of composites, test, design and analysis methods will only be mentioned briefly.

5.2 Constituent materials

5.2.1 Reinforcements

Glass, carbon and aramid fibres are the most widely used reinforcement types in filament winding applications. The filament winding technique has many applications, including the civil, military and aerospace industries. Hence, cheap fibres with inferior properties, as well as very expensive fibre types with superior properties, are used.

Glass roving reinforcements are most widely used in filament winding applications. Glass fibres have types like 'A', 'C', 'D', 'E', 'R', 'S', 'S2', 'E-CR' and 'AR'. Generally, E and S2 type of glass rovings are commonly used in filament winding applications. E-Glass is the most common reinforcement type for the filament winding technique. E-Glass is considered as the standard reinforcement in filament winding, and it has a low alkali content (less than 1%) [1]. E-glass is used widely due to its low cost. Although, S2-Glass has superior mechanical properties compared to E-glass, it has a limited usage due to the higher cost.

105

Carbon and aramid fibres can also be used in filament winding applications. Both aramid and carbon fibres have better mechanical properties than any kind of glass fibres. Additionally, carbon and aramid fibres both have better thermal properties. Aramid fibres are generally used in products which may be subjected to impact loading. A disadvantage of aramid fibres is the ease of moisture absorption, which has a negative effect on the structural performance of the material. Both carbon and aramid fibres have applications in the aerospace and defence industry. Both types of fibres are more expensive than E-glass.

The fibre types are compared according to their mechanical properties in **Table 5.1**, in which the heading 'bonding' refers to fibre-matrix bonding.

Since, in a composite part, reinforcements carry most of the applied load, the mechanical properties of fibres are very important for designers. The use of composite materials keeps increasing, due to their low weight or, in other words, their high strength to weight ratio. Table 5.2 shows average values for some of the mechanical properties of E-glass, S2-glass, carbon and aramid fibres. The values given in this table are approximate, and are not recommended for use in design.

Table 5.1 Fibre property comparison							
Fibre	Strength	Elastic Modulus	Bonding	Abrasion Resistance	Maximum Strain	Density	Cost
E-glass	Medium	Medium	High	High	Medium	High	Low
S2-glass	High	High	Medium	High	Medium	High	Medium
Carbon	High	High	High	Low	Low	Low	High
Aramid	High	High	Low	Low	Med.	Low	High

Table 5.2 Mechanical properties						
Fibre	Density (kg/m³)	Tensile Strength (GPa)	Specific Strength (MPa m³/kg)	Elastic Modulus (GPa)	Specific Modulus (MPa m³/kg)	Maximum Strain (%)
E-glass	2,500–2,600	2.0–3.0	0.8–1.2	65–80	25–30	4.5–5.0
S2-glass	2,400–2,500	3.4–3.8	1.3–1.6	80–110	35–45	5.0–5.5
Carbon	1,700–1,800	3.5–4.0	2.0–2.3	230–250	130–140	1.0–2.0
Aramid	1,400–1,500	2.5–3.0	1.6–2.0	100–140	65–95	2.0–3.8

Ranges are given for each value in **Table 5.2**, because there are some differences among similar products depending on the producer. For example, Akzo Nobel carbon has the strength of 3600 MPa, an elastic modulus of 230 GPa and a density of 1700 kg/m^3; this compares to Tenax Fibres GmbH carbon, which has values of 3950 MPa, 238 GPa and 1770 kg/m^3, respectively.

Generally, the weight of the rovings is measured with a bulk mass unit, which is called 'tex'. This is a unit for expressing linear density, and is equal to the mass (in grams) of 1000 metres of filament. The producer of the composite structure selects the roving with the most appropriate tex. The rovings may have tex values of 400, 600, 800, 1200, 2400, and 4800, etc. The diameters of the fibre bundles increase with the increasing tex value.

Fibres are coated with certain materials for various reasons. This coating is called sizing, and it is a very important factor in composites manufacturing technology. Film-forming organics and polymers are used to protect the reinforcement during processing. Adhesion promoters are applied to improve the mechanical properties and moisture resistance of the composite system. Chemical modifiers react to form a protective coating. Although it is rarely used commercially, an interlayer type of sizing is used to enhance composite properties. Detailed information about functions, types, chemical composition and theory of sizing can be found in [2].

The most widely used sizing are silanes, which are generally applied with a film-forming polymer. Silanes are used with polyester and epoxy resin systems, which are the most common resin types used in filament winding applications. This type of sizing increases the fibre-matrix bonding and protects the fibres against moisture. More information about possible interpenetration of silane and epoxy molecules can be found in [3]. A more detailed discussion of reinforcing materials is presented in Chapter 2 of this handbook.

5.2.2 Resin systems

As in other composite part production techniques, a resin system is used in filament winding: this resin system is called the matrix material after curing. The purpose of the resin system is virtually identical to the one used in other composite parts produced by different techniques. The characteristics of the resin system are:

* they hold the fibres together and so help to distribute the load evenly

* they protect the fibres from abrasion during winding and from abrasion and corrosion after curing

- they provide interlaminar shear strength

- they help to control the chemical and electrical properties of the composite part

As matrix systems, both thermosets and thermoplastics can be used in the filament winding process. The use of thermoset resin systems is definitely more common and standard.

Epoxy, polyester and vinyl ester thermosets are the most widely used resin systems in filament winding. Phenolics, polyimides and silicones are other types of resin systems used in winding applications. Because of reasons such as their cost, level of experience, difficulty in handling and processibility, their use is limited to special applications such as high temperature and electrical uses.

Epoxy resins are the most common resin systems in high performance filament winding applications. Additionally, these resin systems have a wide range of mechanical and thermal properties. As the result of the superior properties of epoxy resin systems, they are used in aerospace and military applications. Resin systems are obtained after mixing the appropriate catalysts and inhibitors, as defined by the resin producer. The resin producer also describes the mixing ratios, but composite manufacturers may subsequently change these ratios.

Polyester and vinyl ester resin systems have relatively lower costs than epoxy resin systems. For this reason, these resin systems are extensively used in commercial applications. The design and processing characteristics of these resins can easily be adapted to the filament winding technique.

A thorough summary of the properties of resin systems is given in Chapter 2. The resin system selection process for filament winding applications is discussed in this section. Further information concerning epoxy, polyester and vinyl ester resin systems can be found in [4, 5].

The designer determines the required mechanical, thermal or chemical properties of the matrix. Also, a resin system may be selected by the designer. However, manufacturing-dependent parameters may not allow the use of the resin system selected by the designer. The main processing-dependent characteristics of a resin system are the curing temperature, viscosity and pot life.

The curing temperature directly affects the mechanical properties of the composite part produced. Generally, approximate curing cycles are defined by the resin producer, but each composite manufacturer determines the curing cycle for their specific needs. For high temperature applications, the decomposition temperature and the T_g of the resin systems become important. Depending on the application and resin system, the composite part can be cured in ovens or at room temperature. Oven curing generally is preferred for mass production and applications requiring accuracy. The room temperature curing takes much longer and, generally, lower T_g are obtained. Fast and high temperature curing increases the T_g.

Resin viscosity is an extremely important parameter during processing. Low viscosity values are preferred for better wetting of the fibres. However, the resins with very low viscosity values may flow out of the part during winding. To prevent this problem, in general, the resin bath temperature is slightly increased, so that the viscosity of the resin in the bath decreases and wets the fibres well. Unfortunately, after removal from the resin bath, the resin on the fibres has an increased viscosity. If the viscosity is too high, the fibres are unevenly coated with resin. The composite manufacturers consider the 0.25–1.25 Pa-s range as the most suitable for filament winding applications.

Another important parameter is the pot life. After mixing the resin, hardener and, if necessary, the accelerator, the resin system starts to gel, even at room temperature. Thus, a resin system should be selected to ensure that its gel time is longer than the winding time of the part. If gelation occurs during the winding, a non-uniform resin distribution occurs and a weakened structure results. To reduce such problems, continuous control of the resin/curing agent mixture or a controlled batch mixing are preferred.

Temperature-viscosity-time relationships for several resin systems are provided by the producers, but laboratory testing is recommended for more precise control.

Another important factor affecting the final product is the shrinkage. Epoxy resin systems shrink least during curing, with values of 1%–5%, depending on the catalyst and curing cycle. The highest shrinkage is seen in the polyester systems, where values are 2%–8%. Low shrinkage causes lower internal stresses in the manufactured part and provides ease in mandrel removal.

5.3 Manufacturing

Even though the filament winding technique used may be wet or dry, winding methods are very similar. As discussed in Section 5.1, this process involves winding the continuous fibre around a mandrel.

The employment of the filament winding technique in a manufacturing process requires four main stages:

- Preparation: The filament winding machine and the mandrel are prepared. Fibres are placed on the spools of the winding machine, and the mandrel is placed on the spindle of the machine. Mould release agents are applied to the mandrel. If necessary, the mandrel can be heated.

109

- Winding: Fibres are wound onto the mandrel at pre-designated angles and in patterns. Generally, in this part of the process, commercially available computer codes are used to obtain the desired accuracy.

- Curing: After the winding is completed, the part with the mandrel is placed in an oven for curing. The part remains in the oven for a particular time at one or more temperatures. In some cases, depending on the resin system used, the part is left at room temperature for curing.

- Mandrel removal: After the curing process is completed, the mandrel is removed by various methods. The method used depends on the type of the mandrel: they can be multi-use or disposable.

The complete process is highly dependent on the filament winding machine and the winding programme used. In order to achieve the requisite winding angle and winding pattern, several, very similar methods are used.

5.3.1 Winding methods

Polar and helical winding are the two main processes. Each of these processes is used to produce a distinctive filament pattern. Polar winding (sometimes called planar winding), is where the mandrel remains stationary while the fibre feed arm rotates about the longitudinal axis, inclined at the prescribed winding angle. The mandrel revolves one fibre bandwidth at each rotation of the fibre feed arm. This pattern is described as a single circuit polar wrap in **Figure 5.1** [5]. In this case, the fibre bands are laid adjacent to each other until a complete layer is obtained. Such a layer is composed of two plies oriented in positive and negative winding angle ($\pm \alpha$) at the end of one complete revolution of the mandrel.

If the mandrel rotates continuously and the fibre feed carriage moves back and forth along the mandrel axis, the process is called helical winding. **Figure 5.2** illustrates the latter, and a filament winding machine is shown. The desired winding angle is obtained by arranging the mandrel rotation and the fibre feed carriage speeds. In this case, after the initial loops, the fibre bands are not adjacent and several loops are required to cover the mandrel and obtain the first layer.

As the derivatives of basic polar and helical modes, a few types of windings are also worth mentioning. When winding angles in helical winding approach 90°, the winding type is called hoop winding. In hoop winding, there is one bandwidth per one axial movement of the feed carriage. When the winding angle is very low and the angle is limited with the polar openings of the mandrel, the winding type is called longitudinal.

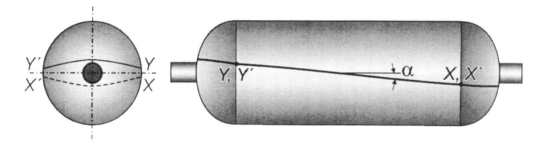

Figure 5.1 Polar winding

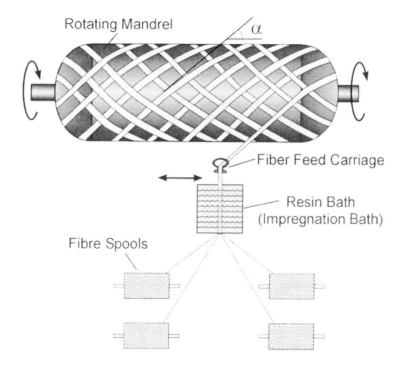

Figure 5.2 Filament winding machine diagram and helical winding

Winding type is called combination winding, if the helical windings with various winding angles and hoop windings are all applied together in the same winding pattern.

Winding patterns used to be determined by trial and error through winding machine adjustments. Alternatively, they were calculated by manipulation of the geometry in the early filament winding technology. However, with modern technology, winding methods are also a part of the design and, for this purpose, computer-aided design tools are extensively used.

From the designer's viewpoint, designing optimal layers is not the only issue: they must also be reproducible. A production technique is needed to provide accurate fibre placement. Due to the fibre path stability requirements in filament winding, the trajectory of the fibre path and the corresponding winding angles cannot be selected freely. The limitations make it difficult to approach the optimal lay-up and hence complicates the design process. Two other factors directly affecting the winding optimisation are the winding machine itself and the software used.

Fibres are wound onto the mandrel along different paths (or trajectories) which require stability and no slippage. Geodesics are the most commonly used fibre trajectories. Geodesics are curves connecting two points on a surface according to the shortest distance over the wound surface. Fibres placed along geodesic lines do not slip when being pulled, and stability of geodesic winding does not require any friction.

For surfaces of revolution, the geodesic equation can be expressed by the law of Clairaut [6]:

$$r \sin \alpha = R_o \tag{5.1}$$

where

r is the radius at a particular point on the mandrel
α is the winding angle
R_o is a constant

Filaments are not necessarily wound geodesically to be stable. Stable non-geodesic winding, often called semi-geodesic winding, can also be performed. This requires a little deviation from the geodesic paths, depending on the friction required to keep the fibre at the appropriate position.

In recent years, design facilities for filament winding have been developed in tandem with developments in computer technology. Along with powerful computer-aided winding equipment, an advanced design capability has significantly extended filament winding potential and products [6, 7]. Several integrated computer codes have been developed for the design and manufacture of filament wound composite parts, e.g., CADFIL [8],

CADMAC [9], CADFIBER/CADWIND [10] and CAWAR [11]. Quality control tools are also included in some of these codes.

5.3.2 Fibre placement machines and tooling

Two types of machines are designed for polar and helical winding. Each type has variations to add some versatility. The main advantage of the polar machines is that the construction of the feed arm is simpler. Polar machines generally operate with a vertically-positioned mandrel. This eliminates the deflections due to the weight of the mandrel and simplifies the feed arm construction. Usually, inertial effects occur when speeds are changed or movement directions are reversed. Since the rotation of the feed arm is continuous and at a uniform speed, inertial effects are not experienced. The limitation of prepreg materials and the difficulty in the installation of wet winding systems are the main disadvantages of the polar winding machines.

Helical winding machines have two basic movements, which are the rotation of the mandrel, and the traverse of the fibre feed arm. Computer Numerical Controlled (CNC) helical winding machines are currently designed up to 6-axis. These axes are shown in **Figure 5.3**.

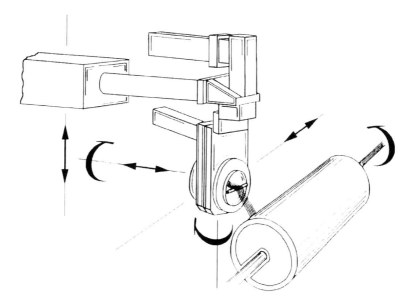

Figure 5.3 The 6-axes of a helical filament winding machine

CNC helical winding machines are used both for wet winding and prepreg winding, and they are the most popular commercially. In order to use 4, 5 or 6-axis types a computer software is usually needed. Generally, axisymmetric parts are manufactured by a filament-winding machine. However, the use of an appropriate computer program, allied to a 4- to 6-axis machine, permits the manufacture of asymmetrical parts (including 'T'-shaped connections and elbows).

Under these conditions, the filament winding ability of a manufacturer is only limited by the size and weight allowed by the machine. Some filament winding machines have multiple spindles and these kinds of machines are suitable for mass production.

With minor mechanical changes, machines which have been designed exclusively for prepreg winding tapes or filament winding can be converted to cater for both types.

Tools used in the filament winding process are rather simple. The most important tool is the mandrel, which is discussed in section 5.3.3. The mandrel should be mounted on the filament winding machines. For this purpose, simple arms and connection parts are used. Another important piece of equipment is the oven, which may be horizontal or vertical. Mandrels are continuously rotated in horizontal-type ovens, or placed vertically in vertical-type ovens. In both cases, in order to heat and cure the wound part evenly, proper air circulation must be provided by using fans inside the ovens. All the ovens should be time- and temperature-controlled.

5.3.3 Winding mandrels

The mandrel design is dependent on the shape of the composite part to be produced. The most important feature of mandrel design is the removal of the mandrel after curing. Mandrel design for open-ended structures like cylindrical or conical parts is relatively easy. A solid or tubular cylindrical mandrel, or a conical steel or aluminium mandrel would be useful for such parts.

For pressure vessels and tanks where the end domes are integrally wound together with the cylinder itself, the mandrel design gains more importance. Mandrel type and material selection are crucial in such a case with respect to ease of mandrel removal. A mandrel must not deflect during winding due its own weight and fibre tension. It should also be able to withstand the elevated temperatures during the curing process. The mandrel types can be summarised as:

Re-useable solid mandrel: Generally, these type of mandrels are composed of metallic materials, and they are appropriate for open-ended part production. Metallic mandrels are the most common type, and have the lowest cost per part.

Collapsible mandrel: Generally, collapsible mandrels are made of segmented metallic parts. Mandrel parts are assembled for winding and, after curing, disassembled for mandrel removal. Collapsible mandrels are rather expensive.

Inflatable mandrel: This type is similar to the collapsible mandrel in terms of usage. Inflatable mandrels are not suitable for resistance to torsional loads. To improve the torque resistance, the mandrel may be filled with materials such as sand. Another method is to perform curing in a closed mould and then apply pressure.

Disposable mandrel: In this case, for each part to be wound, a separate mandrel is produced. After curing, the mandrel is destroyed. Although the material cost is inexpensive, this method is not suitable for the production of large number of parts. The materials used for the disposable mandrel vary. Common examples include low melting point alloys, eutectic salts, soluble plasters and washout sand.

Collapsible mandrels and disposable mandrels all have size limitations. However, by using open end metallic mandrels, huge parts can be wound.

5.3.4 Resin application and curing (wet-dry winding-post impregnation)

The combination of resin with the reinforcement is termed the impregnation process. The impregnation methods are dry winding (prepreg winding, wet rerolled prepreg winding), wet winding and post impregnation.

Prepreg materials offer excellent quality and process control. Although the prepreg material cost is high, using a prepreg winding method produces high performance structures. Also, controlled material storage conditions are required for the raw material. The winding process is similar in all impregnation methods. In prepreg winding, the prepreg material is initially installed on the spools. These prepreg rovings (which are already resin impregnated), are then wound onto the mandrel without any resin bath. To control the tackiness of the material, the spools must be maintained at a certain temperature. Sometimes, solvents and preservatives are used for the same purpose.

In many cases, an autoclave curing may be necessary for high performance products. A controlled volume of resin is applied to a controlled amount of reinforcement in wet rerolled prepreg. The quality control is performed away from the winding process. Pre-impregnated rovings are installed on the spools and the same process in prepreg winding is performed. The pre-impregnated material can be stored in freezers for later use. The cost of this method is relatively lower than for prepreg winding.

If the reinforcing fibres are passed through a resin bath or resin-controlled roller for impregnation, the method is called wet winding. This method is the most common

Table 5.3 Comparison of wet and prepreg winding		
	Wet Winding	Prepreg Winding
Cleanliness	Poor	Good
Fibre availability	Good	Poor
Resin content control	Poor. Depends on many parameters	Good (at constant speed and temperature)
Quality assurance	Poor	Good
Use of complex resin systems	Yes	Very difficult
Resin system data	Available	Available
Simple resin formulation	Necessary	Possible
Fibre damage	Less damage probability	Depends on prepreg producer
Storage	Very easy (long shelf lives)	Freezers must be used or records should be kept.
Winding speed	Low (depends on the fibre quality and winding machine)	High (resin throw during winding is very low)
Room temperature cure	Possible	Not possible
Slippage on non-geodesic surfaces	Low (wet resin causes slippage)	High
Cost	Low	High

commercially. To increase the fibre wetting, the resin bath and/or fibres can be heated to a pre-determined temperature directly before placement in the bath. The control of resin content is difficult and dependent on the resin viscosity, fibre coating, winding speed, fibre tension, bandwidth, etc. The cost of this method is the lowest, but the process is relatively messy since the resin dripping all around and the fact that the resin is cured fast, it requires a lengthy post-cleaning process of the machine and the workplace. A comparison of wet winding and prepreg winding is shown in **Table 5.3**.

The post-impregnation method is similar to the other methods, but in this case, the dry fibres are wound on the mandrel and then the resin is applied. The resin may be applied in a closed mould system and cured in this form. Generally, this method is considered a RTM application. The resin may be applied to the dry wound mandrel manually or the whole mandrel may be dipped in the resin for small parts. The main advantage of post-impregnation is that the fibre slippage during winding is prevented. The cost of this method is lower than dry winding, but post impregnation takes more time.

In all these methods, at the end of winding and resin impregnation, a curing process is required. Depending on the material used, the curing process is performed at elevated temperatures or at room temperature. Curing temperatures are dependent on the resin system used. If the T_g of the resin system is high, then the cure temperature is also high. The wound product may also be cured at low temperatures, but in this case the curing process may even last for years. By keeping the cure temperatures slightly higher than the T_g and increasing the heat transfer rate, the T_g may be increased by a few degrees. This means that the stiffness of the product is slightly increased. The curing temperatures of epoxy resin systems may be even higher than 200 °C.

The complete curing process is in two stages, curing and post curing. Post curing is performed at temperatures around or a little higher than the T_g of the resin system. Conversely, curing is performed at temperatures lower than the T_g.

The most important tool for curing is the oven, which must have time- and temperature-control and be able to maintain a balanced temperature distribution over the product. The polyester resin systems are generally cured at room temperature. However, they may be cured at temperatures of 40–50 °C to shorten the process time.

The application of process control is relatively easy in prepreg winding and relatively difficult in wet winding. However, it is a requirement to be able to produce repetitive parts by using any impregnation and any winding method.

5.3.5 Process control

The quality assurance for composite structures concentrates on the validation of the mechanical, electrical and chemical properties of the cured part. The end product testing results in various testing procedures and publications. Nowadays, quality assurance is initiated a long time before end product testing. [12]. The criteria for quality assurance for composite structure production can be summarised thus:

- Raw material validation

- Material characterisation

- Fabrication, handling, tooling effects

- Cure process control

- Documentation

There are many parameters affecting the quality of the final product manufactured by using the filament winding technique. Controlling all these parameters is crucial in order to produce identical parts at the completion of the process. The rationale of process control is that repeating the same process will result in an identical product. Hence, the cost due to scrap and quality assurance will be reduced. In the wet winding process, the number of parameters affecting the quality of the final product is higher than in other methods. These include:

- Resin viscosity
- Resin bath temperature
- Fibre speed
- Fibre tension
- Surrounding temperature and humidity

Controlling all these parameters is practically impossible. To manufacture a standard product, the process should be repeated almost exactly. In order to detect any change in process parameters, computer-aided quality assurance programmes are used. These systems continuously record the fibre tension, resin bath temperature, fibre speed, fibre length used, surrounding temperature and humidity.

Raw materials are received by the composite manufacturer from the vendor with a certificate. This certificate describes various physical measurements of the raw materials. Nevertheless, a viscosity measurement is usually performed. Thermal analysis tests are carried out to ascertain whether the product is cured. Also, resin/fibre ratio, T_g and resin decomposition temperature are controlled in these investigations.

All the tools must be well-maintained as a part of the process control. Additionally, the human factor cannot be neglected: all the operators must be well-trained and aware of the importance of process control.

The records of all controls and tests are documented and the data periodically compared to the previous set. Therefore, any changes in the process can be detected and explanations of end product property changes determined.

5.4 Mechanical properties

5.4.1 General

Filament wound products are finding many applications, ranging from the defence sector to commercial uses in the automotive and aircraft industries. The defence industry has been using this technology since the 1960s in various applications, including rocket motor cases and missile launchers. The sizes of the latter can vary from small anti-personnel rockets to intercontinental ballistic missiles. Various aerospace and aircraft components utilise this technology. Examples include fuselages and aeroplane wings, as well as parts used in the space shuttle. Also, engine inlet cowls and other structural components in the commercial aircraft industry use this technology. Since the mid-1970s, the automotive industry has placed an increasing emphasis on filament winding applications coupled with metal part integration. High-pressure pipes (sometimes involving highly corrosive fluids) are also a major application area for the filament winding technique.

This wide range of applications necessitates a comprehensive design tool for the structural and failure analysis of these components. This is generally under combined loading, including internal/external pressure, axial loading, torsion and even bending moments. The design procedure should also involve an extensive experimental test programme. This should generate both the basic mechanical properties of the material system used, as well as component level tests to ensure the validity of the computational design study. In sections 5.4.2 and 5.4.3, a general overview of the current design approaches is presented, in addition to the associated test methods at both material and structural level.

5.4.2 Design and analysis methods

The geometry of filament wound structural components is generally in tubular form, with positive curvatures due to the nature of the production method. The majority of applications are with circular cylindrical tubes and consequently the studies on filament wound composite structures are concentrated on such geometries [13–19]. For the filament wound tubes, where metal mandrels requiring extraction after curing are used, tubes have a conical shape. Although this geometry has a slight taper of 0.01%–0.02% in the longitudinal direction, considering the tubes without any taper in the analysis does not pose a problem.

Depending on the application, tubes can be open-ended, as in the case of various pipes and automobile driveshafts, as well as closed-ended tubes for pressure vessel applications. Some of these applications may require the use of composite-to-metal joints, which may further complicate the analysis.

Filament wound and close-ended tubes are subjected to various internal and external loading, as well as inertia loads due to rotational speed (ω).

119

These loads, including:

p_i internal pressure

p_o external pressure

F axial load

T torsion

M bending moment

are shown schematically in **Figure 5.4**. In addition to mechanical loads, the tubes may also be subjected to hygrothermal loads, which have a significant effect by reducing the structural performance of composite systems.

The analysis of filament wound tubes involves the determination of the stress and strain distribution in corresponding layers. This is achieved by using the material properties of the constituent materials. It can also be achieved by using the data provided by material tests of the composite system employed, along with the helix angles of layers. For helical layers, the layer is defined as the combination of plus and minus winding angles since the fibre bands criss-cross by creating a weaving effect. Although this definition creates a difficulty in the determination of material properties of a helical layer, the assumption has been used successfully.

The analysis is mainly conducted by three different approaches. The first method involves the use of thin-walled stress analysis for tubes subjected to pressure, axial, torsion and bending loading [14]. It can be used with good accuracy for tubes with very small thicknesses. It can only be used for preliminary design studies.

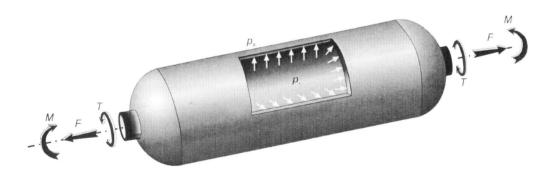

Figure 5.4 Typical loading on filament wound tubes

Another method utilises the shell theory for orthotropic constructions [17, 20, 21]. It is especially helpful for closed tubes where a combination of cylinders with spherical dome ends is used. The approach gives satisfactory results for thin tubes. Thicker tube constructions however, require the use of higher order theories for more accurate results.

The most widely used method utilises a full elasticity solution for cylindrically orthotropic tubes [15, 16, 18, 19]. The method uses originally developed by Lekhnitskii [22] for anisotropic tubes composed of one layer. Later, it was successfully applied to multi-layered structures [14]. Using this method, it is possible to determine the stresses arising from the loads depicted in **Figure 5.4**, including the inertia loading through the application of angular velocity. The method is also applicable to thick-walled sections. The effect of end closures on the stress distribution can not be included in this type of analysis. Indeed, this may lead to an error, especially where end closures are connected to the domes.

In addition to mechanical loads, the hygrothermal effects should also be considered in situations where environmental conditions are a concern. Temperature and humidity are known to affect the structural performance of polymer-based composites [16, 18, 23, 24]. Since filament wound structures are extensively used in moderately severe environments where temperature, moisture and chemicals are involved, this point should be given a special consideration in the design process. The hygrothermal properties of the composite system employed require additional material tests, thereby increasing design costs considerably.

The structures produced by filament winding technology are becoming more complicated in terms of geometry, due to new winding machines with multi-axis and computer-controlled systems. These complicated geometries, coupled with the selection of boundary conditions, necessitate the use of computational methods. The finite element method is extensively used in such studies, including: stress analysis, design optimisation and progressive failure damage determination [23].

Failure and damage mechanisms observed in filament wound composites are factors which complicate the design process. For the failure analysis, global failure theories, where the composite is considered as a homogenous material, are utilised [23]. Although the approach does not consider damage mechanisms at the micro level, it works with considerable accuracy. The number of strength parameters required for a three-dimensional analysis necessitates an extensive material test programme. One problem associated with filament wound structures is leakage, especially in situations where a high internal pressure is concerned. The leakage problem is inherent to polymer-based composite systems. The structure does not fail in the global sense, but leakage through micro cracks in the matrix prevent the load carrying ability of the pressure vessel. Polymer-based or aluminium liners are used to remedy the problem [23].

5.4.3 Test methods

Filament wound structures require an extensive test programme to provide the necessary material and strength data for the design process and to validate the final design. Therefore, it is possible to classify the test programme in two phases. The first phase involves the standard material tests (including coupon tests), to determine mechanical properties and thermal tests for checking the degree of cure, fibre/matrix content and distribution, etc. Tension, shear, three/four-point bending are examples of tests conducted on coupon specimens. These tests are used to determine both the stiffness and strength properties of the material system used. These tests must conform to the standards set by the American Society for Testing and Materials (ASTM).

Another test programme involves structural tests conducted on the tubular specimens. Internal pressure, torsion, bending loads are considered, coupled with different boundary conditions to determine the mechanical performance of the structure, in addition to evaluating the failure and damage mechanisms.

In addition to these destructive tests (which cannot be applied per item), there are a number of non-destructive test methods available. Although these were originally developed for metals, new techniques are emerging which are applicable to filament wound systems [25].

5.5 Current situation and future perspectives

Although a relatively an old process, the filament winding technique has been developed continuously over the years. These developments are in many areas, such as new resin systems, process automation techniques, mandrels and curing methods, as well as computer-controlled filament winders. In addition to these improvements, the ability of the method to adapt to swift changes in advanced composite technology has created many new application areas. Considerable reductions in the process cost, increased product quality, better structural properties and repeatability in manufacturing are the main motivations of emerging applications for the filament winding technique.

Thermoplastic resin systems present a viable option for the future of the filament winding technique. They offer many advantages over thermoset systems because of their reprocessability, repairability and high toughness. Thermoplastics have been used in some filament winding applications. However they have process cost limitations when used on modern filament winding technology. If, in the future, material costs are lowered and the process accelerated with increased levels of automation, thermoplastics like PEEK and PPS may have advantages over thermoset systems [13].

Further progress in the technique is towards integrated process methods. In these systems, the whole manufacturing process (from the sketch board to the final product) can be achieved by using new integrated production and simulation systems. This includes mandrel and product design, the simulation of winding over the mandrel geometry, actual winding and curing. The *in situ* process and curing control is normally part of these systems. These would considerably increase the production speed as well as the product quality.

Currently, the void formation due to improper wetting (or bonding in dry winding systems) and bubbles in the liquid resin are major problems in the process. The mechanical properties of the final product are adversely affected by this feature, and that leads to a low quality product. New compaction systems, coupled with fibre laying operations have been emerging. These reduce the void content, enhance the mechanical properties, provide better bonding, reduce fibre wrinkling and eliminate resin scraping, which sometimes needs hand labour. The application of compaction would also make it possible to obtain a more homogenous resin distribution and layer thicknesses that are translated into a better quality product.

The drive for increasing the speed and automation of the process also created a search for new curing methods. Some new technologies like microwave, ultrasonic energy and electron beam have been used and abandoned for various reasons. However, laser-directed energy is considered to be a viable alternative, especially to the processing of thermoplastic resin systems. Another alternative method is the application of induction heating, which can be used with many resin systems. These alternative curing methods are more applicable to *in situ* winding processes.

Single-use mandrels are important for cases where the product geometry does not allow for the extraction of the mandrel. Plasters and soluble mandrels made of eutectic salt paste and ultrafine sand are some examples of this type of mandrel currently in use. In spite of the many problems associated with soluble mandrels, there is a need for high temperature curing resin systems, since they cannot be used at temperatures beyond 150 °C. The use of ultrafine sand with sodium silicate as the binder was shown to withstand temperatures up to 340 °C, which is applicable to the curing of polyimide resin systems [13].

Finally, the filament winding process has been shown to be a viable alternative method in the production of advanced composites by the introduction of new automation techniques and developments in the material science. It has developed new inventions and applications in related technologies.

References

1. R.M. Mayer and N.L. Hancox, *Design Data for Reinforced Plastics, A Guide for Engineers and Designers*, Chapman and Hall, London, UK, 1994.

2. W.D. Bascom in *Engineered Materials Handbook*, Ed., T.J. Reinhart, ASM International, Ohio, 1987, Volume 1, 122.

3. K. Hoh, H. Ishida and J.L. Koenig in *Composite Interfaces*, Eds., H. Ishida and J.L. Koenig, Elsevier, Amsterdam, 1986, 251.

4. T.W. McCarvill in *Engineered Materials Handbook*, Ed., T.J. Reinhart, ASM International, Ohio, 1987, Volume 1, 135.

5. A.M. Shibley in *Handbook of Composites*, Ed., G. Lubin, Van Nostrand Reinhold Company Inc., New York, 1982, 449.

6. M. Lossie and H. Van Brussel, *Composites Manufacturing*, 1994, 5, 1, 5.

7. S.T. Peters, W.D. Humprey and R.F. Foral, *Filament Winding Structure Fabrication*, SAMPE Publications, Covina, USA, 1991.

8. V. Middleton, M.J. Owen, D.G. Elliman and M. Shearing, Presented at the 2nd International Conference on Automated Composites, Noordwijkerhout, The Netherlands, 1988, 10.

9. G.M. Wells and G.C. Eckold, Presented at the 1st International Conference on Automated Composites, Nottingham, UK, 1986, 6.

10. G. Menges and M. Effing, Presented at the 2nd International Conference on Automated Composites, Noordwijkerhout, The Netherlands, 1988, 11.

11. J. Scholliers, and H. Van Brussel, *Composites Manufacturing*, 1994, 5, 1, 15.

12. R.J. Hinrichs in *Engineered Materials Handbook*, Ed., T. J. Reinhart, ASM International, Ohio, 1987, Volume 1, 729.

13. M.M. Schwartz, *Composite Materials: Processing, Fabrication, and Applications, Volume II*, Prentice Hall, Inc., New Jersey, USA, 1996.

14. S.W. Tsai, *Introduction to Composite Materials*, Technomic Publications, Conneticut, USA, 1980.

15. C. Cazeneuve, P. Joguet, J.C. Maile and C. Oytana, *Composites*, 1992, **23**, 6, 415.

16. C. Wüthrich, *Composites*, 1992, **23**, 6, 407.

17. P.M. Wild and G.W. Vickers, *Composites Part A: Applied Science and Manufacturing*, 1997, **28A**, 1, 47.

18. L.P. Kollar, M.J. Paterson and G.S. Springer, *International Journal of Solid Structures*, 1992, **29**, 12, 1519.

19. J.C. Prucz, J. D'Acquisto and J.E. Smith, *Journal of Pressure Vessel Technology*, 1991, **113**, 86.

20. F.G. Yuan, *Journal of Composite Materials*, 1995, **29**, 7, 903.

21. A. Argento, *Journal of Composite Materials*, 1993, **27**, 18, 1722.

22. S.G. Lekhnitskii, *Theory of Elasticity of an Anisotropic Body*, Mir Publishers, Moscow, Soviet Union, 1981.

23. C.T. Herakovich, *Mechanics of Fibrous Composites*, John Wiley & Sons, Inc., New York, USA, 1998.

24. Y.I. Dimitrienko, *Composites Part A: Applied Science and Manufacturing*, 1997, **28A**, 5, 463.

25. M.M. Schwartz, *Composite Materials: Properties, Nondestructive Testing, and Repair*, Volume 1, Prentice Hall, Inc., New Jersey, USA, 1996.

6 Pultrusion and Other Shaping Processes

M. Giordano, A. Borzacchiello and L. Nicolais

6.1 Introduction

The manufacture of composite materials is based on technologies where the incidence of labour on the total cost is very high. This is one of the limiting factors in the expansion of the composite market in low-cost/high-volume industrial sectors such as electrical, construction and automotive. Pultrusion is the only continuous, highly-automated composite manufacturing process offering high-volume and good-quality production. These are some of the reasons for the market growth of the pultrusion market, and **Figure 6.1** depicts such growth in North America and Europe [1–8].

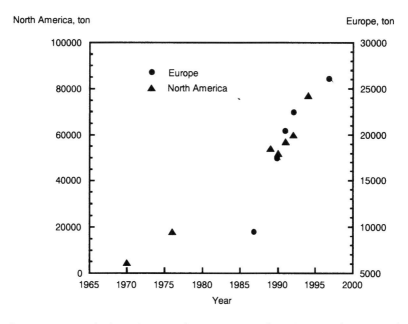

Figure 6.1 North American and European pultrusion market growth

Some of the advantages of using pultruded polymer-based composites are: high specific strength, chemical and corrosion resistance, machinability, adhesive and mechanical joining, dimensional stability, non-magnetic properties and electromagnetic transparency. For many years, production was concentrated on small, simple shapes for the electrical market, due to low electrical conductivity and high environmental resistance properties. This market is relatively mature, but not stagnant. Applications include: electrical enclosures, antennae and guy rods, lighting poles, case tray systems, transmission poles and pylons. In the construction market, pultruded materials are used for both semi-structural and structural applications, such as window frames, cables, rods and bars, grids and meshes, beams and columns, panels and plates. Previously, the growth in sales of standard structural profiles has been limited by the lack of knowledge amongst the construction community about the range and properties of available pultruded profiles. However, today it is the fast growing industrial sector both in Europe (10% annual growth) and in US (15% annual growth). Recent processing technology advancements have extended the applicability of the pultrusion process beyond the traditional constant shape/section profiles used for the majority of commercially pultruded products. Pullshaping and pullforming expanded the market with the production of non-linear and non-constant cross section parts. At the same time, the possibility of pultruding with a wide range of reinforcements (E-glass, S-glass, carbon, Kevlar) and matrices (polyester, epoxy, phenolic, thermoplastic) opened new markets in aerospace, consumer and recreation, transportation and medical sectors [9–15].

6.1.1 Process description

Pultrusion is a continuous composite manufacturing technology that integrates the reinforcement impregnation and the composite consolidation in the same process. The process consists of several successive units: the reinforcement supplier, the impregnation of the reinforcement with liquid resin, the preforming, the consolidation die, the pulling system and the sawing unit. The process is based on the transformation of the liquid resin in the solid final composite while it passes through the pultrusion die. A solid-like state must be reached at the end of the die, allowing the pulling system to act. During pultrusion, fibres in the form of tape, woven, and/or mat are driven through a resin impregnation system. After the resin excess is removed in preforming guides, the fibre/resin system acquires the desired shape and passes through the cure process in a heated die, acting as a continuous reactor. Usually, different heating zones are present along the die, and these are dependent on several factors, such as the type of resin, the pulling speed, and the length of the die. The application of the dragging force to the pultruded parts is performed using a pulling device which is able to impose the desired processing speed. Finally, the pultruder is equipped with a sawing system to cut off the continuous composite produced. **Figure 6.2** illustrates the pultrusion process and the various components are subsequently explained.

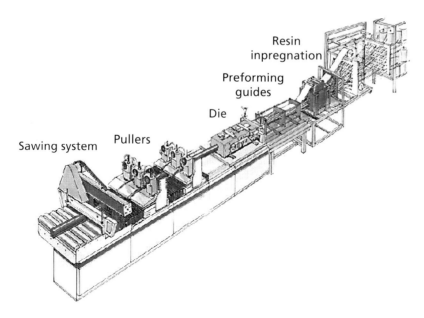

Figure 6.2 The pultrusion process scheme

- Reinforcement supplier

Rovings are used for unidirectional pultrusion, or fabrics may be used to add off-axis fibres. The large number of strands may produce tangles in the formation, and to avoid this problem the fibres are passed through an alignment card just before entering the resin bath. It also prevents twisting of the roving.

- Resin impregnation - wet bath

Continuous strand roving and mats are pulled into a vat of resin that contains liquid resin, inhibitors, curing agent, colorant, fire retardant and other ingredients. To ensure that the fibres are fully wetted, the strands are passed through a series of rollers, which flatten and spread out the individual rovings. A complete wet-out of the fibres must be achieved by controlling the resin viscosity, the residence time and the mechanical action on the reinforcements. The wet bath technique for resin impregnation has several drawbacks, especially in the case of phenolic, bismaleimide and epoxy resins. A modification is to directly inject resin into the fibre preform at the pultruder entrance by using a pressurised pumping system. This technique is suitable in the case of resins having a short pot life, allowing for on-line resin mixing.

- Preforming guides

Guides in Teflon, ultra high molecular weight polyethylene (UHMWPE) or steel are used to force the impregnated reinforcement into the desired shape.

- Heated die

The die must maintain fibre alignment, compress the fibres to the desired volume fraction, and cure the composite in a relatively short period of time. The die temperature profile is selected depending on the resin type. Several heating elements are distributed along the length of the die. In particular cases radio frequency heat sources induce a volumetric heating of the sample, no conductive heating occurs, but each part of the sample absorbs energy uniformly.

- Pulling system

The basic requirement of a pulling system is to transmit the maximum available pulling force at various speeds to the profile exiting the die. **Figure 6.3** shows the Caterpillar pulling system, which is able to provide a continuous pulling force.

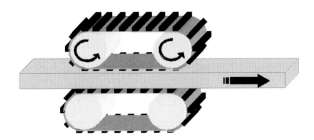

Figure 6.3 Caterpillar pulling system

Conversely, twin gripping-pulling presses with reciprocating pulling action produce a continuous pulling speed. The presses are translated horizontally to grip the profile. In order for the process to be continuous, two presses are used, with one pulling while the other travels back to its initial position. The one gripping-pulling press system makes the pulling action discontinuous.

- Sawing system

A cut off saw is positioned beyond the pulling station. The saw is mounted on a moving table so it can make clean cuts as the pultrusion continues to travel.

6.1.2 Pultrusion modelling

The heart of the pultrusion process is the 'pultrusion die'. It consists of a die with a cavity of the required product cross section and a heating mechanism where the composite consolidation takes place. The variables affecting the behaviour of the material during die passing can be identified as the temperature, the degree of polymerisation, and, consequently, the matrix viscosity, which is a function of the former two. In fact, the processing of thermosetting composites is accompanied by polymerisation reactions (cure process) and rheological changes of the matrix that strongly influence the final properties and the quality of the composite part. The initial stages of the cure process are not only associated with a significant increase of the material viscosity, but are also coupled with a heat generation due to the exothermic thermosetting reactions. The relative rates of heat generation and heat transfer determine the local values of the temperature (and therefore the values of the advancement of the reaction), and of the viscosity (through the thickness of a composite part). In fact, in thick composites, different physical states (namely liquid, rubber, and gelled or ungelled glasses) can be reached at the same time in different sections as a function of temperature and degree of cure. The chemical and physical transformations set the force interactions with the die walls. At the beginning of the die, the liquid resin contributes to the pressure forces. As the viscosity increases, it results in an increasing shear stress at the product/die interface. At the gel point, the product debonds from the die surface and the pulling force is determined by the friction coefficient between the solid composite and the die wall.

6.1.2.1 The thermokinetic and chemorheological properties of thermosets

The characteristics of the curing process and the final properties of thermosetting-based polymeric composites are strongly dependent upon the thermokinetic and chemorheological properties of the matrix. Generally, the cure of a low molecular weight prepolymer initially involves the transformation of a fluid resin into a rubber. This then transforms into a solid glass as a result of the chemical reactions of the active group present in the system, which develop a progressively denser polymeric network. The growth and branching of the polymeric chains are due to intramolecular reactions that initially occur in the liquid state, until a critical degree of branching is reached and an infinite network and an insoluble material is formed. A clear understanding of the chemical and physical transformations is provided by the TTT Gillham diagram, (**Figure 6.4**).

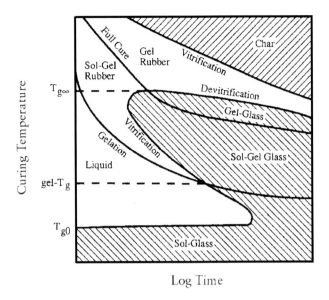

Figure 6.4 The time temperature transformation diagram for a thermoset.

The diagram shows seven zones of the physical status for the material. T_{g0} is the glass transition temperature of uncured resin and no reactivity could be assumed below this temperature. $T_{g\infty}$ is the glass transition temperature of fully cured resin and above it the resin could gel without vitrification. Gel-T_g is the temperature at which the gelation and the vitrification happen simultaneously. The resin passes different phases depending on the curing temperature. For curing temperature below gel-T_g, the resin vitrifies without gelling. For temperatures between gel-T_g and $T_{g\infty}$, the system (which initially is a liquid), gels and, as time increases, vitrifies. The T_g is the most relevant parameter of a polymer. As described by the TTT diagram, the relationship between the evolving T_g and the conversion (α) are crucial in studying the cure phenomena: they determine the solidification occurrence and the ability to produce a gelled polymer or not. For a large number of thermoset systems, the following semi-theoretical relationship [16] holds (Equation 6.1):

$$\ln(T_g) = \frac{(1-\alpha)\ln(T_{g0}) + k\alpha\ln(T_{g\infty})}{(1-\alpha) + k\alpha} \tag{6.1}$$

where $k = \Delta Cp_\infty / \Delta Cp_0$ (ΔCp being the variation of specific heat between the liquid/ rubbery state and glassy state, respectively).

(a) Thermokinetics

DSC is extensively used in the determination of kinetic parameters for thermosets. The DSC technique measures the rate of the heat generated during the cure. The assumption is that the heat flow is proportional to the reaction rate, $d\alpha/dt$ (T, α). Models based on DSC data provide no information on the kinetic mechanisms of the reaction and are suitable for industrial applications. **Table 6.1** shows the models used for several thermosetting systems and the literature sources.

Table 6.1 Kinetic models of thermosetting systems		
Reaction Rate Expression $d\alpha/dt$(T, α)	Thermoset System	Literature Source
$K(T)(1-\alpha)$	Epoxy (DGEBA/DCA)	[17]
$K(T)(1-\alpha)^2$	Epoxy (DGEBA/Amines)	[18, 19]
$K(T)(1-\alpha)^n$	Polyester, Epoxy (Novolac)	[22]
$(K_1(T)+K_2(T)\alpha^m)(1-\alpha)^n$	Polyester,	[23]
	Epoxy (Novolac/filler),	[24]
	Epoxy (DGEBA/Amines)	[25]
$(K1(T)+K_2(T)\alpha^m)(1-\alpha)(B-\alpha)$	Epoxy (TGDDM/DDS)	[26]
where K has the usual Arrhenius form of K(T) = ($k_0 e^{E/RT}$)		

(b) Chemorheology

Chemorheology studies the effect of the curing reaction and temperature on the viscosity of the resin, $\eta(T,\alpha)$. An increase of the temperature will reduce the viscosity of the resin, whereas viscosity will increase as the polymer network develops. Models have been proposed that account for the conversion effect α, directly or through the glassy transition evolution T_g (α) during curing. **Table 6.2** shows the models used for several thermosetting systems and the literature sources.

Table 6.2 Chemorheological models of thermoset systems		
Viscosity Expression, $\eta(T.\alpha)$	**Thermoset System**	**Literature Source**
$\eta_\infty \exp\left(\dfrac{A}{T} + B\alpha\right)$	Epoxy Polyester (Vinylester) Epoxy (Fiberite)	[27] [27] [28]
$\eta_\infty K(T)\left(\dfrac{\alpha_{gel}}{\alpha - \alpha_{gel}}\right)^n$	Epoxy (Hercules 8552) Epoxy (Ciba Geigy 6376) Polyester (Isophtalie)	[29] [30] [31]
$\eta_{T_g} \exp\left[\dfrac{c_1(T - T_g(\alpha))}{c_2 + T - T_g(\alpha)}\right]\left(\dfrac{\alpha_{gel}}{\alpha - \alpha_{gel}}\right)^n$	Epoxy (Araldite)	[32]
$\eta_{T_g} \exp\left[\dfrac{c_1(\alpha)(T - T_g(\alpha))}{c_2(\alpha) + T - T_g(\alpha)}\right]$	Epoxy (TGMDA/Novolac/DDS) Epoxy (E-PEK/DDS)	[20] [33]
$\eta_{T_g} \dfrac{\exp\left[\dfrac{c_1(T - T_0)}{c_2 + T - T_0}\right]\exp\left[\dfrac{b_1(T - T_g(\alpha))}{b_2 + T - T_g(\alpha)}\right]}{\exp\left[\dfrac{b_1(T - T_g(\alpha = 0))}{b_2 + T - T_g(\alpha = 0)}\right]}$	Epoxy (TGBA/TMAB)	[34]
$\eta_{T_g}\left(M(\alpha)\right)^{3.4} \dfrac{\exp\left[\dfrac{c_1(T - T_g(\alpha))}{c_2 + T - T_g(\alpha)}\right]}{\exp\left[\dfrac{c_1(T_0 - T_g(\alpha = 0))}{c_2 + T_0 - T_g(\alpha = 0)}\right]}$	Epoxy	[29]

(c) The pultrusion die

The processing conditions, established by the imposed temperature profile, are a significant factor in the performance of a pultrusion die. The parameters that can be controlled during the pultrusion process are the die temperature profile and the pulling speed. The polymerisation reactions are activated by the increment of temperature when the fibre/resin system is driven through the die. As a consequence of the exothermic nature of the reaction, the heat developed can overheat the pultruding parts. This can generate strong gradients of temperature, degree of polymerisation and viscosity across the thickness. Therefore, the wall temperature in the die must be designed in order to control the development of such gradients, also assuring a complete gelification before the composite reaches the final section of the die. Conversely, an early gelification can lead to higher pulling forces and increased probabilities of blocking, degradation, and/or damage of the material. Therefore, another critical parameter is the pulling speed affecting the residence time of the resin into the die as well as the position of the gel point and the final degree of reaction. At the same time, the pulling speed is also responsible for the mass rate that must be maximised for obvious economical reasons. The processing behaviour of the material during pultrusion is affected by the geometry of the part and by the physicochemical characteristics of the composite components. The thickness of the composite affects the heat diffusion characteristics and subsequently the degree of reaction, temperature, and viscosity profiles. Obviously, the choice of the thermosetting matrix system will determine the characteristics of the reaction kinetics and of the chemorheology associated with the polymerisation process [35–44].

(d) Temperature and conversion profiles

Even if complex die design needs a 3D model to carefully establish the process optimisation, the effect of processing parameters can be analysed using simpler 2D models. A useful perspective of the phenomena developing along the pultrusion die can be obtained by modelling the temperature and conversion distribution. The energy balance equation (an example of a rectangular cross section shown in **Figure 6.5**) accounting for the convective heat transfer along the pulling direction z, the heat conduction in the transverse direction x and the heat generation due to the polymerisation reactions is able to give quite general information about the relationship between the operating parameters.

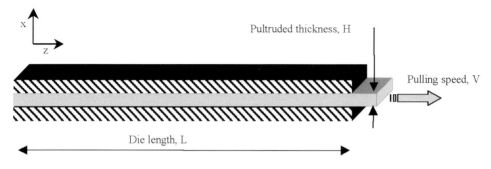

Figure 6.5 Simplified scheme of a pultrusion die

The dimensionless form of the energy balance equation introduces characteristic dimensionless numbers able to characterise the process, that is, (Equation 6.2):

$$\frac{1}{G}\frac{\partial \theta}{\partial \zeta} = De\frac{\partial^2 \theta}{\partial \xi^2} + St\frac{d\alpha}{d\vartheta} \tag{6.2}$$

Table 6.3 defines the symbols used in the energy balance equation.

Table 6.3 Legend of symbols used in the energy balance equation	
$\zeta = z/L$	Dimensionless length
$\xi = x/H$	Dimensionless thickness
$\vartheta = t/t_{gel}$	Dimensionless time
$\theta = \dfrac{T - T_0}{T_{die}^{max} - T_0}$	Dimensionless temperature
$St = \dfrac{H_{reaction}}{c_p\left(T_{die}^{max} - T_0\right)}$	Stefan number (The ratio between the heat generated by chemical reaction and the heat accumulation in the material
$G = \dfrac{L}{V \cdot t_{gel}}$	Gelling number (The ratio between the residence time in the die and the isothermal gel time)
$De = \dfrac{k_x t_{gel}}{\rho c_p H^2}$	Deborah number (The ratio between the heat conduction and the heat accumulation in the slab)

Some conclusions can be drawn by analysing the typical values obtained by the dimensionless numbers in the case of pultrusion. For glass fibre/polyester resin composites, typical Deborah numbers are of the order of 0.1, while for carbon fibre/epoxy laminates they are of the order of 10. These strong differences arise from the higher thermal conductivity of carbon fibres with respect to the glass fibre and the generally higher gelling times of epoxy resins compared with polyester matrices. Typical values of the Stefan number in the case of pultrusion are of the order of unity. The Gelling number indicates the capability of the process to produce a composite characterised by a given conversion. A process presenting a value of the Gelling number less than unity cannot gellify inside the die. On the other hand, a system with a high Gelling number produces fully cured composites.

The temperature, conversion and viscosity profiles can be obtained from the solution of the energy balance equation (Equation 6.2) by specifying the constitutive equations for the materials used in terms of thermal properties, polymerisation kinetics and chemorheological behaviour. The temperature profile along the die or the thermal fluxes, the fibre/matrix inlet temperature, as well as the pulling speed provide the appropriate boundary and initial conditions.

In the following example, the optimisation of the die temperature and pulling speed is shown, based purely on thermal modelling. The die can be divided into different zones at different temperatures and a convective heat flux (post-curing stage) can be provided outside of the die. The processing of an epoxy matrix/carbon fibre-based composite is analysed.

The following criteria, referring to polymer matrix behaviour, can be applied to the optimisation of the pultrusion process:

- a final degree of reaction > 0.8 is required

- the exothermic temperature peak must be minimised to avoid matrix degradation and strong gradients of degree of reaction and viscosity

- the gel point must be reached at the end of the die

- the pulling speed must be maximised in order to produce the highest product rate

Figure 6.6 shows the temperature profiles at the centre and at the skin of the laminate for simulation A.

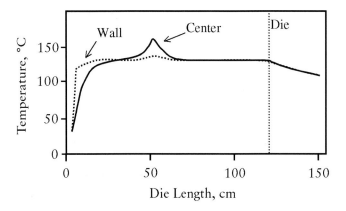

Figure 6.6 Temperature profiles along the die (simulation A)

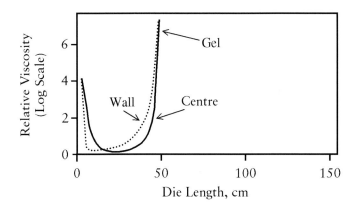

Figure 6.7 Viscosity profiles along the die (simulation A)

At the entrance the cold pultruding element is heated by the mould walls, the mould gives energy to the composite that raises the temperature. The phenomenon is slower within the composite due to conduction. The temperature profiles suggest an inappropriate setting of the operating conditions. Moreover, the gelification, corresponding to a large increase in viscosity (**Figure 6.7**), occurs in the first part of the die as a consequence of a high Gelling number. In this case, either a large part of the pultrusion die is not used for consolidation or higher pulling forces may be required, and blocking, degradation and/ or damage of the material may occur.

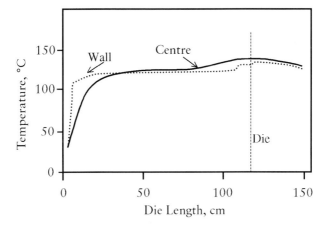

Figure 6.8 Temperature profiles along the die (simulation B)

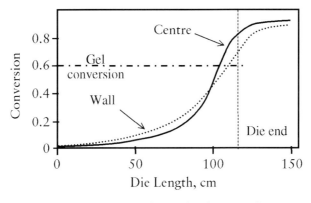

Figure 6.9 Conversion along the die (simulation B)

In this case, the optimisation criteria cannot be fulfilled with an isothermal die, and different heating zones must be provided. Now, since $T_{g\infty} \approx 130$ °C, this means that a temperature of approximately 130 °C is needed to achieve the required degree of conversion. A reduction of the exothermic peak can be obtained by reducing the die temperature in the first part of the pultruder followed by a higher temperature zone to complete the curing. In fact, better results are obtained using two different heating zones, as well as increasing the pulling speed, as shown in **Figure 6.8**.

In the previous case, not only is the temperature peak strongly reduced, but also the temperature gradients along the thickness. As shown in **Figure 6.9**, the cure proceeds uniformly across the composite part and the gel point section, (i.e., the section at which the gelification of the polymeric matrix occurs), is correctly positioned in the last part of the die.

(e) Flow in pultrusion die

Figure 6.10 shows the essential features of material flow through a pultrusion die. Although the reinforcement must have uniform velocity across the channel section and along its length, the resin flows with respect to the reinforcement. The inlet section, where the uncured resin possesses Newtonian behaviour, is typically tapered. In this zone, the resin-impregnated reinforcement carries in the die cavity a resin excess, due to the preforming section. Consequently, a backflow squeezes out the resin, resulting in a pressure increase. A theoretical investigation indicated that the pressure rise is strongly dependent on the shape of the die inlet taper and length of the taper. The linear (conical) taper resulted in the highest pressure rise compared to the round and parabolic tapers.

The viscous flow section is characterised by a viscous layer (of the order of 300 μm between the die wall and the reinforcement/resin mass. Over a small distance from the die wall, the velocity of the resin increase from the zero wall velocity to the pulling speeds passing through a maximum. The dragging of the fibre-resin core induces a shearing effect of the thin resin layer. At the same time, a pressure rise is caused by resin thermal expansion due to resin heating.

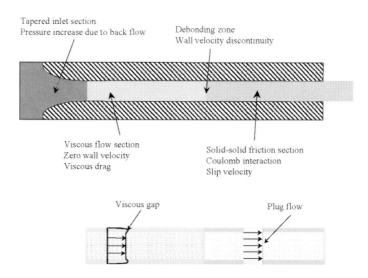

Figure 6.10 Flow phenomena along the pultrusion die

As the resin cures into a gel product, the viscosity increases by several orders of magnitude. The increase in the drag force will result in debonding between the resin and the die wall. A velocity discontinuity exists at the debonding point: the zero wall velocity boundary condition breaks down and the composite moves uniformly. The debonding zone influences the surface quality of the product. The resin may separate from the reinforcement and attach itself to the die surface, resulting in the surface peeling off (sloughing). In the last zone of the pultruder, the composite moves through the die with a plug flow and, consequently, a friction force is generated between the die wall and the solidified composite until the final volumetric shrinkage causes the product release from the surface of the die [45–50].

(f) Pulling force

The pulling force in pultrusion technology is the one of the most relevant parameters. Reduced pulling force allows for the use of smaller, inexpensive manufacturing equipment. Even if an analytical study on the origin of the pulling load is not available, the identification of the various sources can be noted. Earlier studies [51] on the development of wall shear stresses along the die showed that at least three regimes can be identified. **Figure 6.11** reports a qualitative trend for the shear stress along the die length.

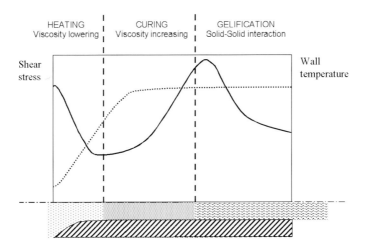

Figure 6.11 Variation of wall shear stress with die axial position

In the tapered inlet and viscous flow sections, contribution to the total pulling force is due to the viscous drag. Initially, shear stresses decrease as resin viscosity is reduced by the rising temperature. Shear stress in the entrance region increases with pulling speed. As the polymerisation proceeds, the increasing viscosity induces higher shear stress. An estimation of the viscous contribution to the pulling force can be calculated from Equation 6.3 [52]:

$$F_{viscous} = 8 \frac{V}{\lambda} C \cdot \int_{0}^{gelation\ length} \eta(z)\ dz \tag{6.3}$$

where the summation is made from the die inlet to the debonding position, assumed to be the gel occurrence; C is the perimeter of the cross section of the pultruded composite, F is the force, V is the pulling speed, λ is the thickness of the viscous layer and η is the viscosity. The thickness of the resin layer should be equal to the distance between the fibres of diameter d_f in an hexagonal array in a composite with a volume fraction of reinforcement ε_f (Equation 6.4), [53]:

$$\lambda = d_f \left[\sqrt{\frac{\pi}{2\varepsilon_f \sqrt{3}}} - 1 \right] \tag{6.4}$$

The relationship between the thermokinetic and chemorheological properties of the resin and the viscous drag relies on the determination of the evolution of the viscosity along the die wall.

After the debonding point (almost gelation), the resin solidification prevents resin shearing and the drag is caused by Coulomb friction ($F_{Coulomb}$) between the solidified composite and the die wall. The contribution of the solid-solid friction to the pulling force is determined by the friction coefficient (and by the pressure profile along the plug flow section, Equation 6.5), [46–48]. Polyester moving on a steel substrate has a friction coefficient of $\mu = 0.19$. Alternatively, for polyester on a glass substrate, $\mu = 0.5$–0.7.

$$F_{Coulomb} = \mu C \cdot \int_{gelation\ length}^{end\ die} p(z)\ dz \tag{6.5}$$

Volumetric shrinkage caused by the polymerisation reaction prevents increase of the pressure due to thermal expansion, and eventually separates the composite from the die wall. The pressure profile and consequently the Coulomb drag arise from a balance of the expansion/shrinkage effect due to the heating and curing of the composite. Therefore, after the gel point, the pressure (P) change can be expressed in terms of averaged temperature (T) and conversion (α) as [48]:

$$\frac{dP}{dz} = \frac{\gamma}{\beta}\overline{T} - \frac{\nu}{\beta}\overline{\alpha}$$

(6.6)

where γ is the thermal expansion coefficient, ν the curing shrinkage coefficient and β the compressibility constant.

Finally, another contribution to the total pulling force arises by the collimation force due to upstream resistance, i.e., the creels, guides and the impregnation system.

6.2 Thermoplastic pultrusion

Thermoplastic pultrusion is a promising continuous manufacturing process, interest in which was aroused in the mid-1980s and quickly grew in the early 1990s. The high viscosity of thermoplastic polymers is the main reason for the much narrower processing window that is possible for thermoplastics than for thermosets pultrusion. With thermoset systems, the viscosity of the plastic required for impregnating roving is approximately 1 to 10 Pa-s. The melt viscosity of thermoplastics is 2–3 orders of magnitude higher, and it is generally not possible to shift the required viscosity range merely by raising the temperature. One of the main benefits of thermoplastic pultrusion is the fact that the thermoplastic matrix can be re-melted [54]. This means that the pultruded thermoplastic can be postformed or joined, making further processing of profiles considerably easier.

Thermoset and thermoplastic pultrusion processes obviously have many elements in common. However, an analysis of the latter does not include modelling of crosslinking kinetics, but instead adds difficulties such as non-Newtonian matrix flow, matrix melting and solidification, and possibly multiple dies. A typical die system for thermoset pultrusion is a one-piece construction having an entrance taper or radius over 2–5% of its total length, so matrix flow relative to the fibres occurs only in a very limited section of the die and is often neglected in theoretical treatments. However, die systems used in thermoplastic pultrusion commonly consist of at least two dies, the last of which is cooled. The heated die, where matrix flow occurs, tends to be shorter in length than the thermoset counterpart. Also, its cavity is tapered over a much greater portion of the die length to create a back flow of matrix to achieve the desired compaction and consolidation. A schematic process of the pultrusion line, for pre-impregnating fibre bundles, is shown in **Figure 6.12**.

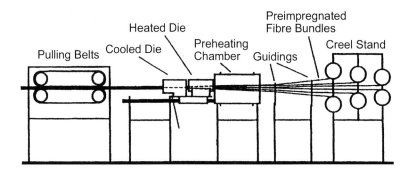

Figure 6.12 Pultrusion line

The development of impregnation technology for thermoplastic matrices has generated considerable interest in the possibility of thermoplastic pultrusion. The potential advantages of thermoplastic matrices over their thermosetting counterparts also include the possibility of rapid processing, improvements in toughness and damage tolerance, chemical stability, environmental resistance and recyclability. Thermoplastic pultrusion offers productivity advantages over conventional thermoset pultrusion, which is severely limited in its operating speed by the rate at which the cure reaction can be induced to take place within the forming die, typical line speeds being of the order of 2 m min^{-1}, depending on the section thickness. Because line speed is, in principle, limited only by the rate at which the pre-impregnating fibre tow can be heated to a temperature above its melting point, and by the rate at which the formed section can be cooled, thermoplastic pultrusion offers potential increases in processing speeds up to values comparable to those encountered in thermoplastics extrusion.

Initially, thermoplastic pultrusion technology was developed using high performance polymers such as PEEK, PEI and PPS. Later on, the technology employed low cost thermoplastic polymers such as polymethylmethacrylate (PMMA), polypropylene (PP), PET and polycarbonate (PC) [55].

6.2.1 Thermoplastic pultrusion modelling

A lot of work has been done on modelling thermoplastic pultrusion. Lee and co-workers [56, 57] developed a model to provide the relationship between the temperature, crystallinity and bonding inside the composite and the required pulling force during pultrusion of fibre reinforced thermoplastic matrix composites. They analysed the pultrusion of thermoplastic matrix composites reinforced by unidirectional fibres. They assumed the cross section to be symmetric with respect to the die's centreline. The die height (H) was constant along the width, which was assumed to be small in comparison to the width (W) and the length (L) of the die. As a consequence of the former geometrical condition, they supposed the fibres to be parallel to the axis and that the variation in temperature and crystallinity exists only in the direction parallel and perpendicular to the die axis (x and z directions). Variations in the width direction were neglected. The pull speed (V) was constant. They wrote the law of conservation of energy equation as:

$$\rho C V \frac{\partial T}{\partial x} = \frac{\partial}{\partial z}\left(K\frac{\partial T}{\partial z}\right) + Q \tag{6.7}$$

where T is the composite's temperature, ρ, C and K are the effective density, effective specific heat and effective thermal conductivities of the composite, respectively. In the same equation, Q is the heat liberated due to the crystallisation, which is zero for amorphous thermoplastics, and for semi-crystalline thermoplastics, it is expressed as:

$$Q = \left(\rho_m v_m H_u \frac{dc}{dt}\right)\left(1 - Vv\right) \tag{6.8}$$

where ρ_m and v_m are the matrix density and the matrix volume fraction, v_v is the void fraction, H_u is the ultimate heat of crystallisation and dc/dt is the rate of change of matrix crystallinity. The expression relating the crystallinity to the temperature and to the cooling rate has to be obtained experimentally.

For the solution of previous equations as boundary conditions, it is assumed that the composite is at a uniform temperature T_0 at the inlet of the die, inside the die and along the surface, and that the heat conducted into the composite is equal to the heat transferred from the die wall. Between the die and the puller, the heat conducted into the composite was assumed to be equal to the heat transferred to the air at ambient temperature. Solutions of previous equations together to the boundary conditions provided the composite temperature and crystallinity as functions of position. They developed a consolidation model taking into account the fact that the surfaces of the layers entering the die are generally irregular and that the layers are not in continuous contact.

To describe the consolidation process, they represented the irregular ply surface as consisting of a series of rectangles. The composite was assumed to have irregularities only in the direction of the width and to be smooth along the axial direction. For the degree of consolidation (the degree of intimate contact), they used the expression:

$$D_c = \frac{a_0/a}{1 + w_0/b_0} \tag{6.9}$$

145

where a_0 and a are the height of each rectangular element at the entrance of the die and inside the die, at position x, respectively, b is the width inside the die at position x, and w_0 is the initial distance between two adjacent elements. Lee and co-workers wrote the conservation of mass for a volume control of width $d\xi$ applied to the i-th layer. This volume control is attached to the material and moves along the die at the constant velocity, V. With the movement of the material along the die, there is a pressure increase on the material, resulting in a lateral motion of the rectangular material and a reduction in the height 'a' of the rectangular element. Lateral flow of the matrix is assumed to be laminar, and the average pressure exerted on the plane which encompasses one unit of rectangle (width $b_0 + w_0$) was derived as:

$$P_{av} = \frac{1}{b_0 + w_0} \int_{-b/2}^{b/2} (P - P_a) d\xi = \frac{\mu_{mf}}{b_0 + w_0} \left(\frac{b}{a}\right)^3 V\left(-\frac{da}{dx}\right) \tag{6.10}$$

where μ_{mf} is the viscosity of the matrix-fibre mixture. At any axial position x, the average pressure is constant across the composite (P_{av} is constant in the z direction). However, the compaction (height a) varies from layer to layer because of the difference in the temperature and thus viscosity. The total change in the composite thickness is given as:

$$2\sum_{i=1}^{N}\left(-\frac{da}{dx}\right)_i = 2\frac{P_{av}}{V}\sum_{i=1}^{N}\left[\frac{b_0 + w_0}{(\mu_{mf})i}\left(\frac{a_i}{b_i}\right)^3\right] \tag{6.11}$$

where N is the total number of layers.

The local composite thickness (at x) was equal to the prescribed local die height and then, Equation 6.12 was derived as:

$$2\sum_{i=1}^{N}\left(\frac{da}{dx}\right)_i = \frac{dH}{dx} \tag{6.12}$$

From equations 6.11 and 6.12, the average pressure can be written as:

$$P_{av} = -\frac{dH/dx}{\dfrac{2}{V}\sum_{i=1}^{N}\left[\dfrac{b_0 + w_0}{(\mu_{mf})i}\left(\dfrac{a_i}{b_i}\right)^3\right]} \tag{6.13}$$

For the consolidation of the layer i, Equation 6.14 can be obtained:

$$\left(\frac{dD_c}{dx}\right)_i = \frac{a_0 b_0}{2} \frac{dH/dx}{\sum_{k=1}^{N}\left[\frac{b_0 + w_0}{(\mu_{mf})_k}\left(\frac{a_k}{b_k}\right)^3\right]} \frac{1}{(\mu_{mf})_i}\left(\frac{a_i}{b_i^3}\right)$$

(6.14)

After integration of equation 6.14 with the boundary condition derived from the fact that at the die inlet $a = a_0$, consolidation of each layer as a function of position x is obtained.

To calculate the force required to pull the composite through the die, four factors are considered:

a) pre-tension applied to the prepeg prior to entering the die, Ft

b) the pressure exerted by the composite on the die

c) Coulomb friction between the solid composite and the die wall

d) hydrodynamic friction resulting from the shearing of the thin fluid resin layer contained between the composite and the die wall.

The total pulling force is the sum of the four contributions and is given by Equation 6.15:

$$F = 2\int_0^L \left[P_{av}(\tan\theta + f_c) + \mu_r \frac{V}{\delta}\right]dx + F_t$$

(6.15)

These three submodels, which were developed separately, were solved simultaneously using numerical methods. The die was composed of two sections, a tapered section with a constant taper, and a straight section 0.25 m in length. The width of the die was constant. The die wall temperature was kept constant at each section. It was assumed that there was no pre-tension on the prepreg and that no resin layer was present between the die wall and the composite. The pulling speed affected both temperature and consolidation distributions. It was found that, for the analysed configuration, the consolidation was complete at the exit of the tapered section in every ply, but the temperature was non-uniform. The composite would have to be brought to a uniform temperature in the straight die section, which follows the tapered section. With respect to the pressure, as the die became narrower, the pressure increased considerably. The pulling force decreased as the taper (θ) increased, which was explained by the fact that for larger angles, the length of the tapered section became shorter. This resulted in a lower pulling force and, moreover for shorter dies, the plies consolidated faster. After consolidation, the heat transfer increased dramatically due to the increase in contact between the plies. This increase in heat transfer resulted in higher temperatures inside the composite and hence, a lower viscosity for the system. The decrease in viscosity further reduced the pulling force.

Other, extensive investigations on thermoplastic pultrusion modelling have been completed by Åström et al., [58–61]. They developed a temperature model in order to predict the temperature of the composite as a function of the distance from the composite centreline and location in the pultrusion line. The composite was schematised as an infinitely long slab of constant height, transversally isotropic. Åström considered heat transfer to occur only through the thickness, and neglected axial heat transfer, viscous dissipation, crystallisation, melting, as well as solidification. As boundary conditions, they considered two cases. Firstly, a constant temperature on the boundary region, which was solved using an analytical model. Secondly, a heat flow from the surfaces was considered which was solved using a numerical model.

For the pressure model, a matrix flow relative to the fibres which occured only in the tapered region of the die was assumed. The matrix flow was only along the length of the fibres, and in the constant cross section it was assumed that a plug flow had developed. Therefore, the composite was analysed as a fibre-filled fluid and as a solid in the constant cross section. The matrix flow distribution in the taper was obtained through superposition of the flow caused by the fibre bed translation through the taper, and the flow caused by the thermal expansion of the composite within the confines of the taper. The flow rate-pressure drop relationship (necessary to calculate the matrix pressure distribution), was written using the Kozeny-Carmen equation [72] modified for flow through aligned beds and considering the shear thinning characteristic of polymeric melts. In particular the Carreau model was used to express the apparent viscosity of the polymer melt as a function of the shear rate. The contribution of the fibre bed to the total pressure load carried by the fibres bed was considered to be negligible at low fibre volume fractions, but important when the fibre bed approaches the maximum packaging.

The full pressure model was used as input to the pulling force model. Åström considered the total pulling force required to pull the composite trough the die as the sum of viscous, compaction and friction resistances. To calculate the viscous resistance, it was assumed that a matrix boundary layer acts as a lubricant between the outermost layer of fibre and the die wall without slipping at both interfaces. For the compaction resistance, the sum of the matrix pressure and of the load carried by the fibre bed were considered, and for a taper with an angle equal to zero, this contribution disappears. For the friction resistance, Åström considered the friction arising from the contact between the bare fibres and the die wall when no lubricating matrix layer was present, and the friction between the solid composite and the die wall of the cooled die before the contraction eliminated the contact. Åström applied the previous model to a pultrusion assembly consisting of two dies, preceded by a preheater not included in the modelling, the first die was composed of a linear taper followed by a constant cross section cavity.

For a 3.2 mm thick glass/PP composite, they found that the process can be run at a pulling speed of up to 16 mm/s, but not at 32 mm/s. The pressure increased with increasing pulling speed and with decreasing composite temperature. The pressure did not depend linearly on the pulling speed, as the matrix was a non-Newtonian fluid. The pulling force decreased with increasing taper angle of linear taper. Åström also compared the modelling results to the experimental results obtained by means of an apparatus for thermoplastic pultrusion. They verified the model results for unidirectional carbon fibres melt impregnated with PEEK (carbon/PEEK) and unidirectional E-Glass fibres melt impregnated with PP (glass/PP). They found that the temperature model predictions were in good agreement with the experimental temperature distribution. A discrepancy was found between the model prediction and experimental data in the tapered region. This was because the model treated the taper as a constant cross section of the same height as the reminder of the die, and considered that the axial heat transfer and heat transfer in the width direction were negligible. The importance of the assumption of negligible axial heat transfer was made evident by the fact that, for the glass fibre system, the discrepancy between the experimental data and the model were virtually absent. In fact, the thermal conductivity of glass fibre was one order of magnitude lower than carbon fibre. With respect to the pressure model, Åström confirmed the qualitative agreement between predictions and experiments, normalising the pressure distribution with respect to the peak pressure. Experimentally, Åström found pressure decay occurring in the beginning of the constant cross-section of the heated die, which was not included in the pressure model results. They postulated that the matrix flow relative to the fibres (not included in the model), occurred. The pressure gradient in the constant cross section of the die will cause matrix flow relative to the fibres, which may explain pressure relaxation. They also found good agreement between the experimental pulling force and prediction of the pulling force model.

Many refinements of the previous basic models have been completed. Hepola et al., [62] adopted a cell model approach. The physical tapered section of the heated die was divided into self-similar cells, i.e., the authors divided the tapered section of the heated die into cells, each one consisting of a tapered region around each fibre. A cell consisted of a tapered region of resin around each fibre. For this geometrical arrangement, they solved the pressure field that yielded the pressure gradient. The pressure profile was integrated over the taper length to determine the pulling force required to overcome this resistance. They described the shear thinning behaviour of the resin using a Carreau viscosity characterisation. To account for the temperature-dependent viscosity and heat generation due to viscous dissipation, they coupled the cell model with the energy equation that describes the heat transfer in the die. The flow in the cell was assumed to be one-dimensional. The geometry of the cell was determined by the fibre volume fraction at the entrance (v_0), and at the exit (v_1), of the taper. As a result of the model, they found that at slow pulling speed, heat flow occurs transversally in the rod and the dissipative term is negligible in the analysis. By increasing the pulling speed, heat transfer will be convective, and viscous dissipation will balance the conduction. The preheater temperature should be maintained close to the actual processing temperature without degrading the polymers. The high preheater temperature lowers and contributes to a homogenous consolidation process in the taper.

Some authors have developed a microscopical solution of the impregnation phenomena inherent to the pultrusion process. Recently, a finite element simulation of the thermoplastic pultrusion process has been developed [63].

6.2.2 Factors that affect thermoplastics pultrusion processes

Many studies have been completed to determine the influence of the processing parameters on the mechanical properties of the thermoplastic pultruded composites, such as flexural modulus, flexural strength, shear strength etc. The most relevant process parameters are: the temperatures of the preheater, the heated and the cooled die and the pulling speed. The effect of material parameters, such as volume of glass fibre content and types of materials (e.g., commingled yarn or preimpregnated tape) have also been studied.

Åström et al. [63, 64] reported the properties of carbon/PEEK composites (with a fibre volume fraction of 60%) pultruded at speeds of up to 10 mm/s. The authors found that the increase in the preheater temperature only slightly increases the flexural strength and modulus. They suggested that an increase in the properties, due to an increase in the preheater temperature, should be much more pronounced for higher pulling speed and/ or thicker layer. This is because the transverse heat transfer may be insufficient, leading to a incomplete molten prepregs and consequently to a very poor ply-to-ply bonding. The increase of the heated die temperature leads to a decrease in the flexural strength and to an unaffected flexural modulus. The strength increases as the die is cooled less: the flexural modulus exhibits a similar, but considerably weaker trend. With an increase in pulling speed a decrease in the flexural strength and modulus has been observed: the decreasing residence time possibly leads to an insufficient consolidation time. An increase in cooling rate (and hence a degree of crystallinity decrease), leads to strength and, to a lesser degree, modulus decrease. The decrease in the mechanical properties with increasing line speed is also reported in other studies [65]. It has been observed that the apparent interlaminar shear strength of Nylon 12/glass composite decreases sharply at low line speeds (0.35–1.00 m/min) and more gradually at higher line speeds (up to 3 m/min), at low applied roll pressure. A roll pressure increase, at low speed, leads to a decrease in shear strength. This may be due to the cooling effect of contact with the roll. Flexural strength decreases with the line speed in a more gradual fashion than shear strength. When increasing the preheater temperature up to 240 °C, the shear and flexural strength appear to remain constant at higher pull speeds (up to 1.5 m/min).

Bechtold et al. [66] studied the properties of pultruded rectangular profiles of commingled glass fibre/PP (GF/PP) yarn with different glass fibre volume fraction and pre-impregnated GF/PP tapes. They found that with commingled yarns, the shear strength decreases slightly with the increasing glass fibre volume content. Up to a glass fibre volume content of 35%, the shear strength was independent of both the forming die temperature and the applied pulling speed. Pre-impregnated tapes showed higher shear strength level but the value decreased rapidly with increasing pulling speed.

Klinkmuller et al. [67] determined the effect of the tapered die cavity angle and pulling speed on the properties of a pultruded PBT powder-impregnated and PBT-surrounded glass fibre bundle system (42% fibre volume fraction). Small angles were shown to create higher pressures at a constant pulling speed. However, at large angles (8.5°, 10°), the pressure increase with increasing pull speed was much smaller than for smaller angles. The density and the flexural modulus showed the highest values at a moderate (5°) angle, and not at extremes of angle size. Shear strength increased with pressure, the flexural modulus was less influenced by the pressure, though pressures lower than 50 MPa were not recommended. For commingled GF/PP composites [68], it was found that the final void content varied between 0 and 2% for speeds of 1 to 5 m/min, and from 1 to 4% for speeds of 5 to 10 m/min. The effect of weathering on the strength of the pultrusion was found to be considerable [69]. Exposure to salt spray was more damaging than exposure to ultraviolet light for GF/PP composites. Maximum damage occurred after exposure to ultraviolet light followed immediately by exposure to salt spray, where the shear strength was reduced by 50%.

6.3 Pullshaping

Pullshaping offers a method for pultruding a continuously-formed, elongated, non-linear composite element. Two arrangements are possible. In one scheme, resin-impregnated reinforcing material is pulled through a preheating zone and then passed through a pair of die frames. These die frames rotate about a main shaft, and carry first and second accurately-shaped die sections, respectively. The pultruding element is moved with the first and second die sections during rotation and through the die cavity where the shape of the die cavity is imparted to the impregnated reinforcing material. A final curing mechanism is associated with the rotatable die frames to cure the matrix as it moves through the die cavity to produce a rigid, non-linear pultruded article. The first and second die frames, initially connected to form a continuous die cavity with the fixed die section, have a reciprocating action. These movable die sections will continuously move in such a way as to form a continuous process [70].

In the second scheme, resin-impregnated reinforcing material is pulled through a preheating zone and then introduced into an initial shaping and forming die. The material is then introduced into an externally-heated consolidation and curing die, which has a die cavity which is curved over its length in the direction of movement of the material. A final cure is imparted to the resin in the reinforcing material in this consolidation and curing die. A pulling mechanism is employed to pull the impregnated reinforcing material through the pre-heating zone and the dies.

6.4 Pullforming

Pullforming is a method for producing a pultrusion product which has a variable cross section by using a specially-adapted, temperature-controllable pultrusion die. The process provides a continuous method for producing pultruded articles of variable cross sections by varying the temperature within the die as the resin/reinforcement mass is pulled to provide an element having one part substantially fully cured and one other part uncured. The uncured part is then reshaped and cured to form the final product, which possesses a variable cross section. The process comprises the steps of pulling a reinforcing material impregnated with a thermoset resin through a temperature-programmable die. The temperature of the pultrusion die and the pulling speed can be set in order to fully cure a predetermined length of material. A rapid lowering of the temperature of the die allows a length of material to pass through the die substantially uncured. The uncured part emerging from the die is reshaped and cured immediately downstream to the die before the cutting zone. **Figure 6.13** shows some typical shapes produced with the pullforming process [71].

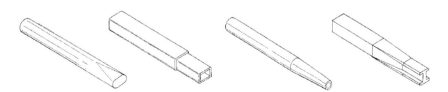

Figure 6.13 Typical pullforming products

6.5 Pullwinding

Pullwinding is a method of producing high performance composite tubes which combines the techniques of conventional pultrusion and continuous filament winding. This technique allows the inclusion of conventional longitudinal reinforcements with helically-wound layers, which provide torsional properties and hoop strength. The technology interfaces a self-contained, free-standing winding unit with a pultrusion machine, feeding hoop or angled fibres between layers of unidirectional fibres for subsequent cure in the pultrusion die. The pullwinding equipment consists of twin winding heads, which revolve in opposite directions over a hollow spindle. The fibre can be angled with the preferred direction. **Figure 6.14** shows the principle of the pullwinding process.

The pultrusion mandrel and impregnated reinforcement pass through the hollow spindle, while the winding heads apply dry fibre over the impregnated fibre. Subsequent layers of impregnated unidirectional can be fed between the winding heads and after the last winding head before the package of reinforcement enters the pultrusion die. The technique permits the production of components for pressure/vacuum applications.

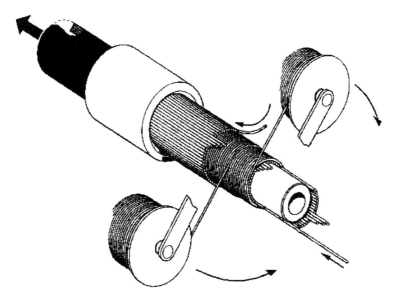

Figure 6.14 Pullwinding process

6.6 Pultruded articles

Pultruded articles are widely used in various markets, including:

Table 6.4 Applications of pultruded articles	
Market	Application
Agriculture and Horticulture	Cloche Systems; Electric Fence Posts; Flower Supports; Garden Stakes; Grape Beater Rods; Greenhouse Structures; Horticultural Stakes; Net Supports; Stock Sticks; Tunnel House Supports
Building	Coving; Concrete Rebars; Hand Railing; Hi-tensile Retainer Wall Pins; Ladders; Lamp Posts; Portable Platforms; Security Fencing; Scaffolding; Structural Beams; Tool Handles
Defence	Camouflage Net Poles; Elements for structural applications (wing, fuselage, etc); Tent Poles
Electrical	Antennas; Cable Ladders and Ducts; Ducting Rod; Fuse Holders; Hot Sticks; Insulators; Line Spacer Bars; Oven Handles; Power Pole Cross Arms; Slot Wedges; Switch Actuators; Transformer Spacer Bars; Optical Fibre Cable Support
Marine	Gaffe Handles; Marker Posts; Mooring Whips/Pier Ladders; Railing; Rods for Buoys; Sea Wall Protection Units; Stanchions; Walkways; Yachting Battens
Mining	Booms; Oil Sucker Rods; Rock Bolts; Structural Supports
Recreation	Batons; Bicycle Flag Sticks; Bows and Arrows; Diving Spears; Flag Poles; Fishing Rods; Golf Shafts; Hockey Sticks; Kite Struts; Paddle Handles; Sail Battens; Ski Poles; Skifield Barriers and Signposts; Sports Field Marker; Tent Poles; Xylophone Tone Bars
Sewerage and Water Supply	Filter Grids; Grating; Hand Railing; Ladders; Outlet Capping Bolts; Tank Bracing; Sludge Scrappers; Sluice Gate Guides; Sections for Treatment Plants
Transport	Dunnage Bars; Handrails; Highway Delineator Posts; Material Handling Components; Seating; Sign Posts; Taut Liner Components
Miscellaneous	Bed Slats; Handles; Refrigeration Components; Seating; Standard Structural Shapes; Surveying Pegs; Tubes for Jigs and Fixtures; Umbrella Components; Vibrator Suspensions

6.7 Other shaping processes

6.7.1 Centrifugal casting

In centrifugal casting, mats or pre-impregnated components are manually positioned within a hollow, cylindrical metal mould, which is usually located inside an oven. Catalysed resin is sprayed onto the mat as the mould slowly revolves. Alternatively, roving may be chopped within, and placed inside, the walls of the open-ended mould. The oven door is closed and the mould, while being heated to 82–93 °C, is rotated at a high speed. Centrifugal force distributes and compacts the resin and the reinforcement uniformly against the inside walls, prior to resin cure. The mould is then stopped and the part is removed. Good surfaces (both inside and outside) are usually obtained. The mechanical properties of the parts produced by this method are in general lower than those produced by filament winding. By centrifugal casting, it is possible to obtain high production rates with automation. Large diameter, dished-bottom tanks are preferably made by this method.

6.7.2 Tube rolling

In the tube rolling technique, precut lengths of a prepreg are rolled onto a removable mandrel by a number of techniques. Some of the various tube rolling methods are illustrated in **Figure 6.15**.

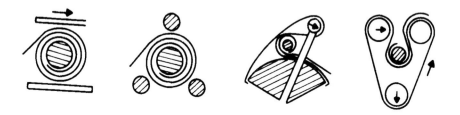

Figure 6.15 Schematics of various tube rolling methods

In this method, the uncured tube is wrapped with a heat-shrinkable film or sleeve and is cured at high temperatures in an air-circulated oven. As the outer wrap shrinks tightly on the rolled prepreg, air is squeezed out at the ends. After the curing, the mandrel is removed and a hollow tube is formed. The tube rolling process is more suitable for simple lay-ups containing 0 and 90° layers. Its advantages are low tooling costs, simple operation, good control of resin content and resin distribution, and fast production rates.

6.7.3 Continuous laminating

In the continuous laminating process, roving is chopped onto a film of resin which has been doctored onto, and is supported by, a cellophane or other suitable carrier sheet. The sheet is initially passed through a kneading device to eliminate entrapped air, then is covered with a second sheet and passed through squeeze rollers to establish a closely controlled, finished panel thickness. Finally, the laminate passes through a curing oven (93–149 °C), which may contain shaping rollers for corrugations. Panels are then stripped of the carried sheets and cut to length. In continuous electrical sheet manufacture, mat of fabric passes through a bath of resin followed by a roller application of carrier sheets and curing (**Figure 6.16**).

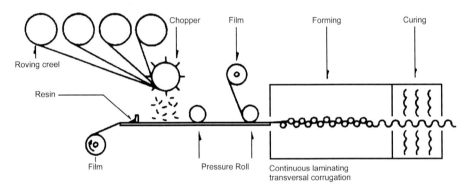

Figure 6.16 A continuous laminating line

6.7.4 Blow moulding

The basic blow moulding process involves the following steps:

- producing firstly a plastics parison or preform (tube, pipe or test tube shape)

- placing this parison or preform into a closed two-plate mould (the cavity in the mould represents the outside shape of the part to be produced)

- injection of air into the heated parison

- allowing air to be blown against the mould cavity

- permit cooling of the expanded parison

- open the mould to remove the rigid, blow moulded part

Blow moulding techniques can be divided into three major categories: extrusion, injection and stretch blow. Extrusion blow moulding principally uses an unsupported parison, while injection blow moulding uses a parison supported on a metal core pin. Stretch blow moulding can start with either the extrusion or blow moulding process.

Blow moulding offers a number of technical and economical advantages in the manufacture of plastic items, particularly when compared to other manufacturing processes, such as injection moulding. These advantages are: the possibility of re-entrant curves (irregular), low stresses, possibility of variable wall thicknesses and favourable cost factors.

6.7.5 Extrusion

In the extrusion process, a granulate is converted to a high viscosity paste by heat and pressure. The basic components of the extruder are shown in **Figure 6.17**.

In this technique, plastic pellets are fed into the feed throat from the hopper. As the unmelted resin exits the feed throat, it comes into contact with a rotating screw. The extruder screw is contained within a heated barrel, which is maintained at a higher temperature than the screw. Since the plastic tends to adhere to hotter surfaces, the unmelted materials will stick to the barrel and slide onto the screw. As the plastic moves forward, it is heated by the barrel in addition to the frictional heat generated by the compounding and compression actions of the screw. By the time the resin leaves the screw and the barrel, it has been melted, compressed and mixed into a homogeneous melt.

Short fibre composites can be fabricated into useful parts by extrusion and moulding.

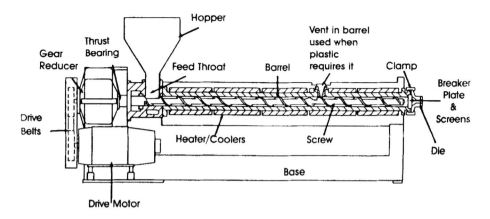

Figure 6.17 A typical single screw extruder with continuous melt flow

6.7.6 Sandwich foam coextrusion

The sandwich foam coextrusion process is a combination of the film and the foam coextrusion processes. In producing sandwiched foam products, the core-forming polymer B (containing a blowing agent), is coextruded with the skin-forming polymer A. A large number of combinations of polymer systems may be used for the skin and core components of a sandwiched foam. In the selection of materials, both core-forming polymer B and skin-forming polymer A can be identical (except that B contains a blowing agent), or they can be different polymers.

References

1. G.F.C. Barrett, S. Johnson and P. Pontrandolfo, *Poliplasti e Plastici Rinfozati*, 1992, **40**, 416/417, 34.

2. H. Lass, *Chemical Week*, 1986, **138**, 14, 34.

3. *Reinforced Plastics*, 1993, **37**, 10, 40.

4. G. Barrett and S. Johnson, *Reinforced Plastics*, 1992, **36**, 7/8, 20.

5. V. Buseyne, *Reinforced Plastics*, 1992, **36**, 7/8, 26.

6. J. Ketel, *Reinforced Plastics*, 1990, **34**, 11, 22.

7. D. Smock, *Plastics World*, 1989, **47**, 3, 12.

8. *Reinforced Plastics*, 1991, **35**, 11, 44.

9. *Materials Edge*, 1992, **36**, 6.

10. F. Wunderley, *Reinforced Plastics*, 1991, **35**, 11, 32.

11. L. Carrino and G. Caprino, *Macplas*, 1993, 18, **148**, 70.

12. A. Weaver, *Reinforced Plastics*, 1996, **40**, 5, 30.

13. M.C. Gabriele, *Plastics Technology*, 1995, **41**, 3, 36.

14. W.C. Magill, *Reinforced Plastics*, 1994, **38**, 10, 26.

15. G. Caprino, *Poliplasti e Plastici Rinforzati*, 1992, **40**, 415, 54.

16. R.A. Venditti, J.K. Gillham, *Journal of Applied Polymer Science*, 1997, **64**, 1, 3.

17. G.C. Martin, A.V. Tungare, B.W. Fuller and J.T. Gotro, Presented at the 47th Annual SPE Conference, ANTEC 89, New York, NY, USA, 1989, 1079.

18. M.E. Ryan, *Polymer Engineering and Science*, 1984, **24**, 9, 698.

19. M. Arellano, P. Velázquez and V.M. González-Romero, Presented at the 47th Annual SPE Conference, ANTEC 89, New York, NY, USA, 1989, 838.

20. P.L. Chiou and A. Letton, *Polymer*, 1992, **33**, 18, 3925.

21. M.E. Ryan, MSc Thesis, (Engineering), McGill University, Montreal, Canada, 1973.

22. T.H. Hsieh and A.C. Su, *Journal of Applied Polymer Science*, 1992, **44**, 1, 165.

23. C.D. Han and K.W. Lem, *Polymer Engineering and Science*, 1984, **24**, 7, 473.

24. A. Hale, M. Garcia, C.W. Macosko and L.T. Manzione, Presented at the 47th Annual SPE Conference, ANTEC 89, New York, NY, USA, 1989, 796.

25. A. Dutta and M.E. Ryan, *Journal of Applied Polymer Science*, 1979, **24**, 3, 635.

26. J.W. Lane and R.K. Khattak, Presented at the 45th Annual SPE Conference, ANTEC 87, Los Angeles, CA, USA, 1987, 982.

27. W.I. Lee, A.C. Loos, G. Springer, *Journal of Composite Materials*, 1982, **16**, 6, 510.

28. M.R. Dusi, W.I. Lee, P.R. Ciriscioli and G.S. Springer, *Journal of Composite Materials*, 1987, **21**, 3, 243.

29. J.M. Kenny, A. Apicella and L. Nicolais, *Polymer Engineering and Science*, 1989, **29**, 15, 973.

30. J.M. Kenny, A. Maffezzoli and L. Nicolais, *Composites Science and Technology*, 1990, **38**, 4, 339.

31. S.M. Moschiar, M.M. Reboredo, J.M. Kenny and A. Vasquez, *Polymer Composites*, 1996, **17**, 3, 478.

32. M. Opalicki and J.M. Kenny, First annual report research project Brite/Euram no. BREU-CT91-0451, 1991.

33. Q. Wang, T. He, P. Xia, T. Chen and B. Huang, *Journal of Applied Polymer Science*, 1997, **66**, 4, 799.

34. D. Hesekamp and M.H. Pahl, *Rheologica Acta*, 1996, **35**, 4, 321.

35. E.P. Calius, S-Y. Lee and G.S. Springer, *Journal of Composite Materials*, 1990, **24**, 12, 1299.

36. W.X. Zukas and N. Tessier, Presented at the 32nd International SAMPE Symposium, Anaheim, CA, USA, 1987, 1288.

37. C. Chen and C.M. Ma, Presented at the 38th International SAMPE Symposium, Anaheim, CA, USA, 1993, 1291.

38. R.M. Hackett and S-Z. Zhu, *Journal of Reinforced Plastics and Composites*, 1992, **11**, 12, 1322.

39. L. Aylward, C. Douglas and D. Roylance, *Polymer Process Engineering*, 1985, **3**, 3, 247.

40. C.D. Han, D.S. Lee and H.B. Chin, *Polymer Engineering and Science*, 1986, **26**, 6, 393.

41. E. Lackey and J.G. Vaughan, *Journal of Reinforced Plastics and Composites*, 1994, **13**, 3, 189.

42. J.P. Fanucci, S. Nolet and S. McCarthy in *Advanced Composites Manufacturing*, Ed., T.G. Gutowski, John Wiley & Sons, New York, 1997, 259.

43. G.L. Batch and C.W. Macosko, *AIChE Journal*, 1993, **39**, 7, 1228.

44. A. Trivisano, A. Maffezzoli, J.M. Kenny and L. Nicolais, *Advances in Polymer Technology*, 1990, **10**, 4, 251.

45. S.M. Walsh, M. Charamchi, Presented at the ASME 25th National Heat Transfer Conference, Anaheim, CA, USA, 1988, 23.

46. R. Gorthala, J.A. Roux and J.G. Vaughan, *Journal of Composite Materials*, 1994, **28**, 6, 486.

47. R. Gorthala, J.A. Roux and J.G. Vaughan, *Journal of Reinforced Plastics and Composites*, 1994, **13**, 6, 484.

48. S.M. Moschiar, M.M. Reboredo, H. Larrondo and A. Vasquez, *Polymer Composites*, 1996, **17**, 6, 850.

49. D-H. Kim, P-G. Han, G-H. Jin and W.I. Lee, *Journal of Composite Materials*, 1997, **31**, 20, 2105.

50. H.L. Price and S.G. Cupschalk in *Polymer Blends and Composites in Multiphase Systems*, Ed., C.D. Han, American Chemical Society, Washington, DC, 1984.

51. A.G. Gibson, C.Y. Lo, D.W. Lamb and J.A. Quinn, *Plastics and Rubber Processing and Applications*, 1989, **12**, 4, 191.

52. M. Giordano and L. Nicolais, *Polymer Composites*, 1997, **18**, 6, 681.

53. M.A. Bibbo and T.G. Gutowski, Presented at the 44th Annual SPE Technical Conference, ANTEC 86, Boston, MA, USA, 1986, 1430.

54. W. Michaeli and D. Jürss, *Composites Part A: Applied Science and Manufacturing*, 1996, **27A**, 1, 3.

55. M.L. Wilson and J.D. Buckley, *Journal of Reinforced Plastics and Composites*, 1994, **13**, 10, 927.

56. W.I. Lee, G.S. Springer and F.N. Smith, *Journal of Composite Materials*, 1991, **25**, 12, 1632.

57. W.I. Lee, G.S. Springer and F.N. Smith, Presented at the 36th International SAMPE Symposium, San Diego, CA, USA, 1996, 1309.

58. B.T. Åström and R.B. Pipes, *SAMPE Quarterly*, 1991, **22**, 4, 55.

59. B.T. Åström, *Composites Manufacturing*, 1992, **3**, 3, 189.

60. B.T. Åström and R.B. Pipes, *Polymer Composites*, 1993, **14**, 3, 173.

61. B.T. Åström and R.B. Pipes, *Polymer Composites*, 1993, **14**, 3, 184.

62. P.J. Hepola, S.G. Advani and R.B. Pipes, Presented at the 25th International SAMPE Technical Conference, Philadelphia, USA, 1993, 736.

63. S.M. Haffner, K. Friedrich, P.J. Hogg and J.J.C. Busfield, *Composites Science & Technology*, 1998, **58**, 8, 1371.

64. B.T. Åström, P.H. Larsson, P.J. Hepola and R.B. Pipes, *Composites*, 1994, **25**, 8, 814.

65. B.J. Devlin, M.D. Williams, J.A. Quinn and A.G. Gibson, *Composites Manufacturing*, 1991, **2**, 3/4, 203.

66. G. Bechtold, R. Reinicke and K. Friedrich, Presented at the 8th European Conference on Composite Materials (ECCM-8), Naples, Italy, 1998, Volume 2, 585.

67. V. Klinkmüeller and K. Friedrich, *High Technology Composites in Modern Applications*, Eds., S.A. Paipetis and A.G. Youtsos, University of Patras, Greece, 1995, 214.

68. A. Miller and A.G. Gibson, Presented at the 11th International Conference on Composite Materials (ICCM-11), Gold Coast, Australia, 1997, 139.

69. W.J. Tomlinson and J.R. Holland, *Journal of Materials Science Letters*, 1994, **13**, 9, 675.

70. W.R. Goldsworthy, E.E. Hardesty and H.E. Karlson, inventors; Goldsworthy Engineering Inc., assignee; US Patent 3,873,399, 1975.

71. J.E. Sumerak, inventor; no assignee; US Patent 5,556,496, 1996.

72. G.P. Carman, *Transactions of the Institute of Chemical Engineers* (London), 1937, 15, 150.

7 Machining and Joining Process

F. Elaldi

The recent introduction of composite materials in aircraft structures has resulted in lower weight and improved performance. These improvements require the development of new design concepts and specialised manufacturing techniques. In addition to weight savings of up to 30%–35% compared to the metal equivalents, composite structures have the advantage of part count reductions of up to 60% over common metal components.

Together with the very high mechanical properties of composite materials, tooling, part removal, cutting, drilling and joining processes begin to become important during the whole manufacturing process.

For the reduction of the cutting, drilling and trimming efforts, optimised design and tooling technologies have to be utilised at the beginning of the manufacturing process. Conventional solid-tool methods or currently preferred new machining techniques are being used to cut, drill and trim cured composite materials. Because of the nature and kind of composite materials, solid cutting and drilling tools should be selected carefully, otherwise inaccurate tooling may cause damage such as delamination, fibre fraying in work piece or premature dulling of the working tool [1].

Water-jet and laser cutting are relatively new techniques, but both are commonly used methods in composite manufacturing.

For the joining of sub-assemblies, mechanical and adhesive bonding techniques are required. There are two basic methods for joining components. They are mechanical and adhesive bonding techniques. Bolt and pin fastening for the mechanical joining technique and co-curing for the adhesive bonding technique will be discussed in this chapter.

7.1 Cutting and trimming

Unique tools and techniques are required to trim and machine composite materials. Specialised cutting equipment is required when working with graphite composites that have layers of metallic materials (aluminum, titanium, etc.), interspersed among the non-metallic layers, as is quite common in hybrid aircraft structures.

The general goals of proper trimming and hole machining operations are:

- no splintering or delamination of surfaces that can be detected by visual examination,

- surface finish of approximately 250 Roughness Average (Ra),

- no discolouration due to overheating,

- loose surface fibres, if any, originating from the hole boundary do not exceed 25% of the hole diameter in length,

- loose fibres, if any, and splinters, do not extend beyond the surface ply.

To meet these types of requirements, specialised tooling is required to provide controlled feeds and speeds.

Tables 7.1 to 7.3 list some types of drills and countersinks that have been found to be effective for drilling composites. **Figures 7.1 to 7.3** show representative samples of specialised cutters designed for composite structures.

Table 7.1 Drill material information						
Material to be drilled	Drill Type					
	Jobbers		Extension	Snake		Comb/ Drill CSK*
	Conventional	Core		Conventional	Core	
	Drill Material**					
Laminate only	1, 2	3	3	1, 2, 3	3	2
Laminate and Aluminum or Titanium	1, 2	3	3	1, 3	3	2
*Combination Drill/CSK: combining drilling and countersunk hole **Drill Material: 1: Carbide Tipped 2: Solid Carbide 3: Cobalt High Speed Steel						

Table 7.2 Point type information for various drills			
Drill Type	C-2 Carbide Tip	Solid Carbide	Cobalt HSS
Jobbers Conventional	NAS 907 Point P-5	T302N7 Burr Point T302N12	NAS 907 Point P-5
Jobbers Core	-	USCTI Stds 14° to 16° clear	USCTI Stds 14° to 16° clear
Extension	-	-	-
Snake Conventional	T302N6	-	NAS 907 Point P-5
Snake Core	-	-	USCTI Stds 14° to 16° clear
Combination Drill/CSK	T302N5 and Spade T302N15	-	-
Flat Flute Drills	-	T302N8	-
Gun Drill (with a Stack Point)	T302XX	-	-
NAS: National Aircraft Standard. USCTI: United States Cutting Tool Institute			

Table 7.3 Typical drilling and reaming parameters				
Hole Diameter (mm) maximum	Drilling		Reaming (If required) (3)	
	Speed, (rpm)	Feed Rate (1, 2) (s/mm)	Speed (rpm)	Feed Rate (1, 2) (s/mm)
	Laminates			
3.97	3000	0.64 to 1.28	3000	0.64 to 1.28
4.76	3000	0.64 to 1.28	3000	0.64 to 1.28
6.35	3000	0.64 to 1.28	3000	0.64 to 1.28
7.94	1800	0.64 to 1.28	1800	0.64 to 1.28
9.53	1000	0.64 to 1.28	1800	0.64 to 1.28

rpm: revolutions per minute

1. Hydraulic feed control units can be adjusted to obtain the specified feed rate in air.

2. Numbers indicate recommended feed rates for positive and hydraulic feed motors. For hand-fed reaming operations, use a slow, steady feed rate.

3. In lieu of positive feed drill motors operating at the specified speeds and feed rates, hand-fed 500 rpm drill motors may be used with appropriate traveller bushings, (i.e., a bushing that is movable) to maintain alignment.

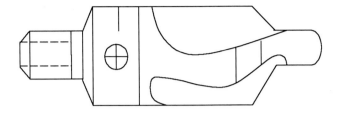

Figure 7.1 Carbide countersink T302Y2 (use with hand countersink cage)

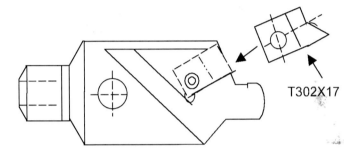

Figure 7.2 Carbide countersink T302Y3 (use with hand countersink cage)

Figure 7.3 Carbide gun drill-T30IN280

For most applications, the drilling of a hole in composite materials will require a two-step operation, **Figure 7.4 and 7.5.**

The plain hole or the countersunk hole is drilled initially, and then a reaming operation follows. When using a gun drill, **Figure 7.6 and 7.7,** a coolant is used to help flush chips from the hole.

Commercial coolants that are specially formulated for these applications are available. When dry-drilling a hole, some type of vacuum system is required to contain the dust generated by the drilling operation [1].

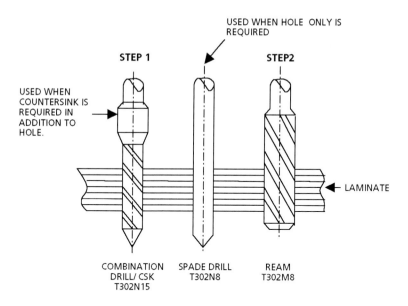

Figure 7.4 Two-step drilling of a laminate

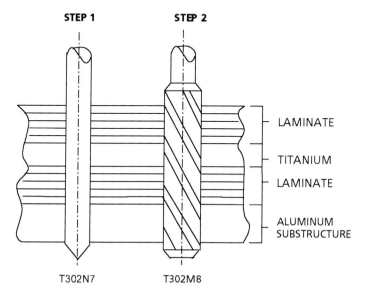

Figure 7.5 Two-step drilling of hybrid structure

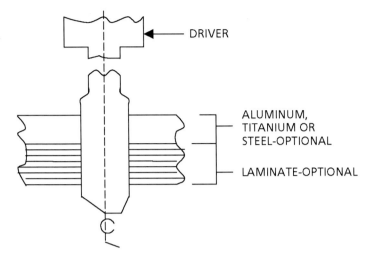

Figure 7.6 Gun drill

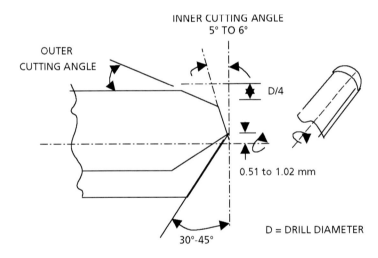

Figure 7.7 Gun drill point configuration 'stack point' for laminates

Because the matrices used in most graphite composites are quite brittle, some type of a back-up on the drill exit side is always required when drilling, to prevent exit side material splintering. Common materials used for back-up are fibreboard, fibreglass laminate, wood or aluminum. When edge trimming is carried out in a router fixture, the router

Table 7.4 Recommended tool, types for the cutting and trimming of composites			
Trimming Operation	Equipment	Cutter Type	Speed Surface m/min
Straight Line Cuts	Radial Arm Saw Table Saw Drill Press Hand Router Disc Sander	Diamond Coated Circular Saw	610 to 3660
	Milling Machine (GR/EP&TI)	Carbide End Mills T-302XX HSS End Mills T-302G6	18 to 27
Irregular Outline	Band saw	Magniband Saw Blade	(portable) 305 to 610 (stationary) 610 to 3660
	Pin Router or Hand Router	Diamond Router	305 to 1980
		Carbide Router Bit	305 to 1980
	Hand Router or Hand Drill Motor	Abrasive Drum	305 to 1980
		Abrasive Disc-Flex Arbor (Mushroom)	122 to 732
Chamfer, DeBurr	Hand	Abrasive Drum	N/A
Finish Operations	Hand	Abrasive Cloth	N/A
GR/EP & TI - Graphite/Epoxy & Titanium; Chamfer - a slope on the edge of a structure; Deburr - the loose fragments after machining.			

fixture itself acts as the back-up material. Even when using the proper drills or cutters with a back-up material, burns and splinters will often occur. The best method for removing these projections is with sandpaper. Generally, a grit of 320 or finer is used [2, 3]. A variety of models of hand power tools exist the for drilling of composites. All of the drill motors have one thing in common, a controlled rate of feed. The nose of the drill motor fastens to the tooling used to locate and clamp the part. When the drill motor is turned on, the drill bit automatically advances at a controlled rate.

The cutting and trimming of laminates requires diamond or carbide saws and router bits. **Table 7.4** lists some of the recommended tool types that are suitable for the cutting and trimming of the composites. When cutting composites, the fibre orientation in the surface ply should be considered to prevent surface splintering. Circular saw blades coated with diamonds are quite effective when used with a radial arm or table saw. Use of these types of tools will help in restricting splintering to a minimum. **Figure 7.8** shows a typical hand-trim setup. Since the rate of feed is not automatic, as in the drilling operation, close attention must be paid to prevent overheating of the laminate.

A fast feed rate will result in a burning odour. A temperature indicator tape, applied to both sides of the saw cut, will indicate a burning condition by a colour change. A proper feed technique will be learnt by the operator through experience, very similar to the manner in which woodworking skills are learnt. As with any form of cutting or trimming of composites, a means of dust collection must be employed.

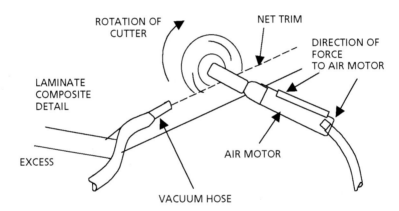

Figure 7.8 Cutter direction for a hand-trim setup

7.1.1 Conventional machining and drilling

Since composites consist of fibre and matrix, they are non-homogenous. Therefore, cutting and drilling tools should be selected very carefully in order to have the desired cutting and drilling surface.

The main problems encountered during cutting and drilling by conventional methods are surface or internal delamination and fibre or resin pullouts.

To reduce the effort for machining, it is common to fabricate the composite structures to almost net shape, (i.e., the final configuration and final dimension of the component after machining). If this is achieved then only the trimming process is required to obtain the net shape. There are several types of cutting tools to cut and trim composites. These are:

• Abrasive tools

• Router cutters and

• Circular blades

The most widely used one is circular blade, since it has a low cost and a long tool life. It is useful for straight and long line cuts. Router cutters become efficient when trimming, thus they are the preferred tools and are also used for trimming and last finishing operations.

In order to fasten two components by mechanical joint techniques (no matter whether they are both composites or not), straight, smooth and high quality holes should be generated for bolts or fasteners. If the holes are wavy, rough and off-the-axis, they may create intensive stress-concentrated areas around the joints and lead to early failures.

Depending on fibre characteristics [4, 5], experiments were performed to optimise the spinning rate and feeding rate of the tools. For instance; up to 40 high qualities, a spade drill/countersink on a 12.7 mm thick graphite reinforced laminate can generate deep holes at a speed of 2800 rpm and a feed rate of 0.040 mm per revolution using spray type of Boelube 70106 coolant at a high pressure [3]. For the aramid-type composite structures it is generally difficult to obtain both good diametrical tolerance and shredding-free surfaces. To prevent shredding and surface delaminations for a clean hole, aramid-type composite structure is preloaded by tensile stress and cut by shear force. Boldt [3] determined experimentally that for a 3.18 mm aramid reinforced composite material, best results were obtained with a spinning rate of 5000 rpm and a feed rate of 0.030 mm per revolution. Sometimes, it may be necessary to drill hybrid structures, e.g., multilayer, mixed carbon and metal sandwiches. There are several methods to drill these kinds of hybrid structures in which peck-drilling (**Figure 7.9**) is one of the most practical methods.

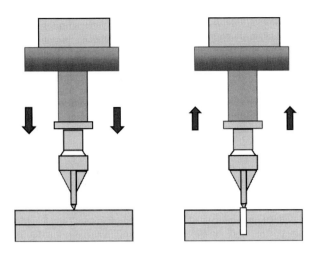

Figure 7.9 Typical peck drilling

7.1.2 Water-jet cutting technique

This method is based on mixing a water-jet stream with abrasives and using the resulting abrasive water-jet to cut non-homogeneous materials (see **Figure 7.10**). The flow rate of the water passing through the nozzle is typically 4 to 8 litres/minute. This creates a pressure at the orifice to provide a jet stream of up to 850 m/s. The other main constituent of the abrasive water-jet technique is the abrasive [6]. After mixing with the jet-stream it removes the material, which exists in the line of cutting axis. Garnet is one of the commonly used abrasive types. The water-jet technique is useful for temperature-sensitive materials. It does not cause burn in the vicinity of cutting edge. It can be initiated at any part of the composite structure. Unlike the conventional techniques, the cut (kerf) is relatively narrow and can be adjusted by altering the water-jet exit nozzle bore size.

The abrasive water-jet method is a new and relatively expensive method. Normally the speed of water and the abrasive are controlled by a computer and for straight and shaped cuttings, a numerical controlled (NC) or computerised numerical control (CNC) controlled X-Y machine, (i.e., a two-dimensional plotter), is required. In order to get a good surface finish of the cut edge, a relatively finer mesh (150 grit) is recommended.

In general, the abrasive water-jet technique is adequate for materials, which have a yield stress of 90 MPa or less. But this technique is more efficient for the denser and thicker materials than the other techniques.

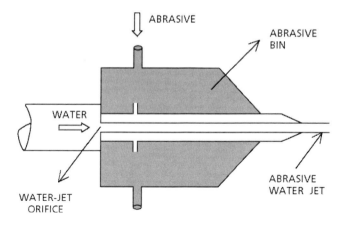

Figure 7.10 Schematic view of water-jet nozzle

7.1.3 Laser cutting

The laser beam cutting method became one of the most popular techniques of the last two decades (**Figure 7.11**). The finish is externally perfect and the laser is generally used for the large pieces. The basic concept is based on concentrating the laser beam on a surface. Laser is a kind of light source, which produces parallel beams, that is why laser can be focused to a certain spot size. The energy that the laser beams carry from the laser source to a surface increases the temperature of that work piece and this energy begins to melt the metals in microseconds or vaporise the organic composite.

Since laser beam cutting is a thermal process, the efficiency can differ from one material to another [7]. The easiest materials to be cut by laser beam are organic composites, such as aramid or epoxy composites. With a CO_2 laser a composite material in 10 mm thickness can be cut easily; an average cutting speed of 6 m per minute can be accomplished for the organic composites. However, if an inorganic material exists in a composite, the cutting speed with a laser will be less than the speed of cutting of organic materials. The main problem of laser cutting for inorganic to organic composites, e.g., carbon-epoxy or fibreglass reinforced plastics, is the melting temperature difference of each of the elements of a composite. For example, the vaporising temperature of carbon is roughly 3600 °C, but the matrix element, e.g., epoxy, can vaporise at a considerably lower temperature than the reinforcement. The strongest component in a composite is the carbon fibre. Hence it is very important to cut the carbon fibres quickly so that the matrix is not exposed to high temperatures for a long time in the cutting line.

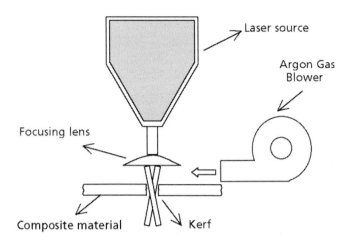

Figure 7.11 Gas-assisted laser cutting

7.2 Joining Techniques

Advanced composites have the potential to produce a weight saving of over 50% of the structural weight of a vehicle. Due to practical constraints, only a 30% reduction is generally realised. One of the major constraints is encountered in the area of joints. One major difference between metallic and composite materials, when working with the latter, is the freedom to design a material with directional properties such as stiffness and strength. However, it can cause difficulties with joints since the composite materials can be highly orthotropic. Also, different stacking sequences and orientations may require complicated joints.

Joints, which must be present when any two components are assembled, are a major source of stress concentrations. Two broad classes of joining techniques are available. They are referred to as mechanical and adhesively bonded joints. Mechanical joints have the advantage that no special surface preparation is needed, they are easy to disassemble, and have no particular inspection requirement. However, the associated holes cause relatively high stress concentrations and the joints can cause a significant weight penalty. Conversely, adhesively bonded joints require surface preparation, are susceptible to environmental effects, are impossible to disassemble and bond strength cannot be predicted and confirmed with a non-destructive inspection method.

7.2.1 Mechanical joint techniques

Mechanical joints require a mechanical fastening agent and are characterised by the cut out (hole) required for the fastener.

Examples are riveted, pinned, and bolted joints. Because of the cut out, only 20 to 50% of the ultimate mechanical strength is generally available for design purposes. Since this is generally intolerable, various local reinforcing methods, such as metallic reinforcements, doublers, (i.e., a build-up), or local ply buildups, are used to strengthen the local bearing or shearing strength for developing acceptable joint strengths.

When fastening two composite structures, the compatibility of these two components is important [8, 9]. In this way corrosion, strength of the components, joining configuration and type of fasteners must be carefully considered.

Some of the advantages of mechanical joints are:

- Utilisation of conventional metal working tools and techniques, as opposed to adhesive bonding procedures,

- Ease of inspection,

- Utility of repeated assembly and dis-assembly for fabrication replacement or repair (accessibility),

- Assurance of structural reliability,

- Less sensitivity to creep,

- High strength to peel loading,

- Easier mating of the surfaces (shape, configuration, etc.)

- Less sensitivity to water, thermal and environmental degradation,

7.2.1.1 Joint geometry

Various kinds of joint geometry have been used in composite structures. There are two main categories, which are single and double shear joints (see **Figure 7.12**). Typical configurations are, straight lap, offset lap, butt lap joint, and tapered butt joints. To reduce the bending stress in the vicinity of the fastener cutouts, the tapered butt-plate joint type is recommended. The possible strength loss due to drilling, the stresses, which may arise from different loading modes and the strength of fasteners have to be taken into account during joint geometry design. In most of the cases, the area of the mechanical joint is reinforced with the same material that we use to produce the joint composite part or component.

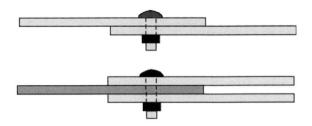

Figure 7.12 Typical single and double shear joints

7.2.1.2 Fasteners for mechanical joints

Choosing the correct type of fasteners for composite joints is based not only on the imposed service loads but also on the galvanic activity (i.e., the degree of chemical activities of different metals) with the parent components as well. For graphite-reinforced composite structures, titanium and stainless steel fasteners, (e.g., A286, a certain type of steel alloy), are widely used. Unless ultra-high fastener strength is required, soft rivets can be used.

Rivets are very common for joining laminates less than 3 mm thick. There are two main types such as solid and hollow, the latter being used when one side access for installation is available. As a general rule, solid rivets produce stronger joints than hollow rivets. In the installation of rivets, some degree of interference between the fastener and the hole can be present. This interference should be well controlled since it could weaken the joints. On external surfaces, countersunk rivets may be required in order to provide a smooth finish and the countersink angle should be as large as possible. Generally, non-countersunk rivets are preferrable to countersunk rivets and washers under the head and tail will enhance the joint strength.

Bolts are also commonly used for single or double sided installation. By tightening the bolts to the laminate, a very high through thickness force will be introduced, hence a very strong joint can be produced. There is no limitation to the thickness that can be joined by using this technique, provided the diameter to thickness ratio is kept within the permitted design strength. As a general rule, bolts with countersunk heads will give weaker joints than those with conventional bolts.

7.2.1.3 Joint failure modes

Failure modes for advanced composite mechanical joints are similar to those for conventional metallic mechanically fastened joints. **Figure 7.13** presents typical

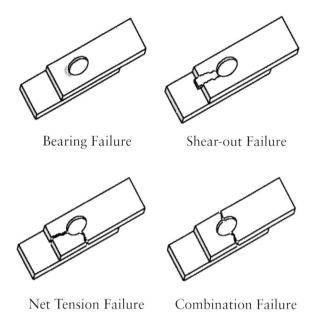

Bearing Failure　　　Shear-out Failure

Net Tension Failure　　Combination Failure

Figure 7.13 Failure modes of composite mechanical joints

simplified presentations of these failure modes, i.e., shear out, net tension, bearing and combined tension and shear out. In addition, bearing or shear failure of the fastener, and bolt pulling through the laminate are other possible failure modes.

Generally, the very wide joints will fail in bearing modes. As the width is reduced, the failure mode will eventually change to tension. However, the width at which the mode will change depends on the lay-up, the hole size effect as well as the basic material properties.

Changing the end distance can have the same effect on shear out failure as width has on tensile failure. Obviously a joint must have adequate end distance if it is to achieve its full bearing strength. The overall tensile strength is reduced as the hole diameter increases. However, for bolted joints in all types of fibre reinforced composite laminate, there is a minimal effect of hole size on net tensile and shear strength. The bearing strength of fibre reinforced composite laminate is similarly unaffected. For low modulus materials such as Kevlar, there is a huge reduction in bearing strength for a diameter to thickness ratio of greater than 3.

7.2.2 Bonded joints

There are two methods for assembling components:

- Co-curing: components are integrally cured to each other in one cure cycle, or

- Secondary bonding: components are cured separately and bonded together with an adhesive film in a secondary operation

Elaldi and others [10] have concluded that the secondary bonding technique is easier to apply than the co-curing technique for complex structures and it usually costs less in tooling due to its simplicity. However, the co-curing technique offers the following advantages:

- Large one-piece structures can be made, thus eliminating joints and discontinuities and improving structural integrity,

- The manufacturing process involves fewer operations, and

- Fewer fit-up problems occur and less sealing is required in assemblies, which reduces costs.

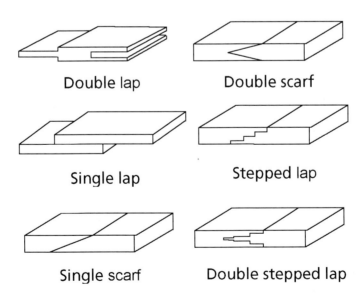

Figure 7.14 Typical joint designs

7.2.2.1 Adhesive materials

Polymeric materials, which structurally resemble the composite matrices are commonly used to adhere two composite structures.

The most commonly used adhesives are epoxies. They could be either one-element (curing agent already mixed in) or two-element (curing agent is mixed at the time the adhesive is needed). A one-element adhesive is generally a sheet-like prepreg without reinforcement. A two-element adhesive is often a paste or liquid. These two types of adhesives could be room temperature or elevated temperature curing systems. For adequate durability, it appears that elevated temperature cured adhesives give better results than the room temperature cured adhesives since their T_g is higher. However, for both variants, the quality of the bond will be dependent on the amount of moisture being absorbed by the composite. Epoxy-based adhesives can provide good bond strength and environmental resistance, but they have very weak peeling strength. The main factors affecting the choice of adhesive systems are the strength, stiffness and ductility.

In addition to epoxy adhesives, acrylic adhesives and thermoplastic materials are also used as adhesives for composites. Thermoplastics are hot melt adhesives for composites. They melt and become fluid when applied to the components to be joined. When they cool, they re-harden again.

7.2.2.2 Curing

The effect of an adhesive bonding depends on the:

- polymeric composition of the adhesive
- surface preparation agents used
- technique of surface preparation
- curing process
- adhesive laying procedure.

To make a strong bond, the surfaces should be cleaned very carefully. All the grease and other contaminants should be removed with special solvents and vapour degreasing carried out by wiping surfaces. After rinsing with water or an alkaline solution, the surfaces are abraded by sanding with a pad, rinsed with water again and dried in an oven if a water rinse is required. The roughened surface has a higher surface energy and area to improve the bonding ability of the composite.

When a composite is bonded to a metal, the metal's surface needs some kind of treatment. Procedures for pre-bonding treatment of aluminum alloys are well established and should be followed for bonding carbon fibre reinforced composite and other polymeric materials. The phosphoric acid anodising technique is commonly used for aluminum cleaning. Titanium alloys, with higher stiffness and strength and a lower coefficient of expansion are better matched to carbon reinforced composites. However, they also need proper surface treatment, particularly if the service environment is in a high humidity atmosphere. Phosphate fluoride is very popular for cleaning the titanium bonding surfaces. The surface preparation of composite materials is less complex than for the metals, the essential requirement is to remove contaminants which can be introduced during the handling or in the manufacturing process. During manufacturing, chosen release methods during moulding of the laminate are in common use. These release systems include release cloth, coated metal sheets or silicone rubbers. It is apparent that hand surface abrasion is generally inadequate and sand blasting with dry alumina grit has proved to be effective.

The two parts to be bonded are joined and kept fixed by clamping after applying the adhesive to the bonding surfaces. This process will ensure that a constant pressure is applied during the curing and thus a constant adhesive thickness can be obtained. The curing process is performed in an oven equipped with a vacuum or in an autoclave or hot press. For good bonding strength the cured adhesive layer should be within 0.1 to 0.2 mm.

Sometimes the co-curing technique is selected according to the compatibility of bonding and bonded components. Semi-cured composite components to be bonded are co-cured with the adhesives. The bond strength is much stronger than the conventional bonding. The other advantage of this method is that there is no need to remove mould release and other contaminants.

The types of adhesive joints (**Figure 7.14**) are similar to mechanical joints [11]. They are single lap, double lap, stepped lap, scarf, single overlay and double overlay. The easiest and cheapest way to bond two components is by single lap since machining is not required.

References

1.   M. Ramulu, M. Faridnia, J.L. Garini and J.E. Jorgensin, *Journal of Engineering Materials and Technology*, 1991, **114**, 113.

2.   H. Hochegn, H.Y. Puw and K.C. Yao, Presented at Machining of Composite Materials Symposium, ASM Materials Week, Chicago, IL, USA, 1992.

3.   J.A. Boldt, *Machining and Drilling of Composite and Composite/Metallic Structures*, Process Specification MA-113, Northrop Corporation, 1985.

4.   G. Lubin, *Handbook of Fiberglass and Advance Composites*, Van Nostrand Reinhold Company, New York, USA, 1969.

5.   Ed., M. Langley, *Carbon Fibers in Engineering*, McGraw-Hill, New York, USA, 1973.

6.   M. Hashish, Presented at the Machining of Composite Materials Symposium, ASM Materials Weeks, Chicago, IL, USA, 1992.

7.   S. Lemma and B. Sheehan, Presented at the Machining of Composite Materials Symposium, ASM Materials Weeks, Chicago, IL, USA, 1992.

8.   Military Handbook, *The Composite Materials Handbook, MIL-17*, Ed., Army Research Laboratories, Technomic Publishing Company, Lancaster, PA, USA, 1999.

9.   D.W. Oplinger in *Fibrous Composites in Structural Design*, Eds., D.W. Oplinger, E.M. Lenoe and J.J. Burke, Plenum Press, New York, USA, 1980.

10. F. Elaldi, S. Lee and R.F. Scott, *Design, Fabrication and Compression Testing of Stiffened Panels for Aircraft Structures*, LTR-ST-1872, NRC, Canada, 1992.

11. L.J. Hart-Smith, *Analysis and Design of Advanced Composite Bonded Joints*, NASA Langley Contractor Report, NASA CR-2218, 1973.

Abbreviations and Acronyms

ACM	Advanced composite material
ARALL	Aramid reinforced aluminium laminate
ARTM	Assisted resin transfer moulding
ASTM	American society for testing and materials
BMC	Bulk moulding compound
BPO	Benzoyl peroxide
C/C	Carbon/carbon
CMC	Carbon matrix composite
CNC	Computer numerically controlled
CTE	Coefficient of thermal expansion
CVD	Chemical vapour disposition
DICY	Dicyanodiamine
DMA	Dynamic mechanical analysis
DMC	Dough moulding compound
DSC	Differential scanning calorimetry
FGR	Fibreglass reinforced
FRC	Fibre reinforced composite
GF/PP	Commingled glass fibre/polypropylene
HBA	p-Hydroxy benzoic acid
HCM	Hybrid composite materials
HIPS	High impact polystyrenehm high modulus
HNA	Hydroxy napthanoic acid
HPLC	High performance liquid chromatography
HT	High tensile
IM	Intermediate modulus
IR	Infra red
LARC	Langley Research Center
LC	Liquid crystalline
LCP	Liquid crystal polymer
LPG	Liquid petroleum gas

MMC	Metal matrix composite
MPD	*m*-phenylenediamine
NASP	National Aerospace Plane of the USA
NDI	Non-destructive inspection
PA	Polyamides
PAI	Polyamide imide
PBT	Polybutylene terepthalate
PC	Polycarbonate
PEEK	Polyetherether ketone
PEI	Polyether imide
PET	Polyethylene terepthalate
PMC	Polymer matrix composite
PMMA	Polymethyl methacrylate
PMR	Polymerisation of monomer reactant
SPP	Polypropylene
PPD	*p*-Phenyleterethalamide
PPPB	*p*-Phenyl bisphenol
PPS	Polyphenylene sulphide
PS	Polysulphone
SPTFE	Polytetrafluoroethylene
rpm	Revolutions per minute
RTM	Resin transfer moulding
SMC	Sheet moulding compounds
SRIM	Structural reaction injection moulding
TERTM	Thermal expansion resin transfer moulding
T_g	Glass transition temperature
TGA	Thermogravimetric analysis
TGMDA	Tetraglycidyl methylene dianine
TMA	Themomechanical analysis
TMC	Thick moulding compound
TME	Thermoplastic polyesters
TPA	Terephthalic acid
TTT	Time-temperature transformation
UHMWPE	Ultra high molecular weight polyethylene

Contributors

Guneri Akovali
Department of Polymer Science and Technology
Middle East Technical University
06531 Ankara
Turkey

Tigmur Agkül
Chief, Material & Process Technologies, Design & Development Department
Tusas Aerospace Industries Incorporated
PO Box 18 Kavaklydere
06692 Ankara
Turkey

Luigi Nicolais
Department of Materials and Production Engineering
University of Naples Federico II
Piazzale Tecchio 80
80125 Napoli
Italy

Michele Giordano
Department of Materials and Production Engineering
University of Naples Federico II
Piazzale Tecchio 80
80125 Napoli
Italy

Assunta Borzacchiello
Department of Materials and Production Engineering
University of Naples Federico II
Piazzale Tecchio 80
80125 Napoli
Italy

Levaend Parnas
Department of Mechanical Engineering
Middle East Technical University
06531 Ankara
Turkey

Nurseli Uyanik
Department of Chemistry
Istanbul Technical University
80626 Maslak
Istanbul
Turkey

Faruk Elaldi
K.K. Teknik Daire Bsk. Ligi
TLFC Technical Department
06100 Ankara
Turkey

Cevdet Kaynak
Department of Mettalurgy and Materials Engineering
Faculty of Engineering
Middle East Technical University
06531 Ankara
Turkey

Samuel Kenig
Director
Israel Plastics and Rubber Center
Josepho Building
Technicon City
Haifa 32000
Israel

Emin Selcuk Ardiç
TAI (Turkish Aerospace Industries)
PO Box 18
Kavaklidere 06692
Ankara
Turkey

Name Index

Main Index